ABOUT ISLAND PRESS

ISLAND PRESS, a nonprofit organization, publishes, markets, and distributes the most advanced thinking on the conservation of our natural resources—books about soil, land, water, forests, wildlife, and hazardous and toxic wastes. These books are practical tools used by public officials, business and industry leaders, natural resource managers, and concerned citizens working to solve both local and global resource problems.

Founded in 1978, Island Press reorganized in 1984 to meet the increasing demand for substantive books on all resource-related issues. Island Press publishes and distributes under its own imprint and offers these services to other nonprofit organizations.

Funding to support Island Press is provided by The Mary Reynolds Babcock Foundation, The Ford Foundation, The George Gund Foundation, The William and Flora Hewlett Foundation, The Joyce Foundation, The J. M. Kaplan Fund, The John D. and Catherine T. MacArthur Foundation, The Andrew W. Mellon Foundation, Northwest Area Foundation, The Jessie Smith Noyes Foundation, The J. N. Pew, Jr. Charitable Trust, The Rockefeller Brothers Fund, The Florence and John Schumann Foundation, and The Tides Foundation.

ABOUT *NEWSDAY*

Newsday, founded in 1940, serves Nassau and Suffolk counties on Long Island. A separate edition, *New York Newsday*, circulates in New York City. Together, they comprise America's sixth largest local newspaper. *Newsday*'s series *The Rush to Burn* was published in 1987 and shared the Worth Bingham award with another series from *Newsday*.

RUSH TO BURN

FROM

Newsday

RUSH TO BURN

Solving America's Garbage Crisis?

ISLAND PRESS

Washington, D.C. □ *Covelo, California*

Cover design: Ben Santora

This Island Press book was originally published as a special reprint, *The Rush to Burn*, © 1988 and 1989, Newsday Inc., Long Island, New York. This edition includes some updating and minor revision.

Library of Congress Cataloging-in-Publication Data

Rush to burn: solving America's garbage crisis? / from Newsday.
p.   cm.
Includes index.
ISBN 1-55963-000-0 (pbk.): $14.95 — ISBN 1-55963-001-9: $22.95
1. Incineration—United States.   2. Refuse and refuse disposal—
United States.   I. Newsday (Hempstead, N.Y.)
TD796.R86   1989
363.7'28'0973—dc19   89-1939
                                                             CIP

Printed on recycled, acid-free paper

Manufactured in the United States of America

10 9 8 7 6 5 4 3

73528

CONTENTS

PART VI
GARBAGE AS GOLD
179

PART VII
SOLUTIONS?
203

PART VIII
EPILOGUE
241

PREFACE

By Anthony J. Marro
Editor, *Newsday*

One day in March 1987, Rex Smith, then a Long Island reporter and more recently *Newsday*'s Albany bureau chief, called the office to pass along a tip he had picked up while working his beat. A barge carrying trash from Islip, Long Island, had just been kicked out of Morehead City, North Carolina, where it had tried to unload, and was drifting around, looking for a new place to dock.

The barge, as the world came to know, was called the *Mobro 4000*. It was being towed by a tugboat named *Break of Dawn* that was commanded by Capt. Duffy St. Pierre. Before its odyssey ended, it had wandered about for 164 days, its 3,186 tons of commercial garbage spurned by officials in North Carolina, Louisiana, Florida, Mexico, Belize, and the Bahamas, as well as by the small town of Sardinia, in upstate New York.

In virtually every port, it was met by officials with court orders and by throngs of reporters. It became the subject both of joke and controversy, of litigation and accusation, and of a fierce public debate that fell somewhere below the level of *The Federalist Papers*.

"If she wants to get porky about it, we'll identify how much of it is hers and we'll leave it on the dock for her," said Frank Jones, the Islip Town supervisor. This was after Queens Borough President Claire Shulman refused to let the barge unload in Long Island City.

"He's an idiot," Shulman replied.

Like most papers, *Newsday* chased the barge and the story,

with Shirley Perlman acting as lead reporter. Perlman was the
first reporter to locate the barge, spent more time with it than any
other reporter, and was on board the tug *Break of Dawn* when it
finally passed under the Verrazano Bridge on its return to New
York.

The day before, trying to locate the tug and the barge in a
storm, the boat that had been chartered by Perlman and a photographer was on its way back to the New Jersey shore when Perlman—who had been monitoring Channel 16 on the marine radio—suddenly heard St. Pierre's voice.

"What are you doing here?" he asked, when she broke in on his
conversation.

"Where you go, I go, kid," she said. "I'm following you to New
York."

By the time Perlman returned to *Newsday*, the Garbage Project, as it came to be called, had been launched—a major effort to
explore and explain all the problems and issues that lay behind
the crisis that the garbage barge had come to symbolize. From a
small group built around two reporters (Tom Maier and Mark
McIntyre), a photographer (Audrey Tiernan), and an editor (Joe
Demma), the project team grew to seven, then a dozen, and then
more. Reporters were drawn from the Long Island and New York
staffs, from the Washington Bureau, from *Newsday* bureaus in
Mexico, London, and Tokyo. Toward the end, there were about 25
people involved in the project, working in a separate office, without windows or clocks, set apart from the main news room.

Questionnaires were sent to more than 200 incinerator plant
operators and builders, and to officials in every state; the results
then were fed into a computer for analysis. The team did a computer study of campaign contributions from the garbage industry
to public officials. It crafted a massive Freedom of Information
Act request, to all Long Island towns, to New York City, and to
the federal government, to obtain information about consultant
contracts. It then hired a consultant of its own to do a study of the
environmental impact statements for all planned incinerators in
the region.

As the project moved along, the large room housing the team
itself took on some aspects of a landfill—a cluttered and surely
unsanitary workplace, where desks were piled high with memos

and reports, where files were jammed with documents on the workings of electrostatic precipitators and reverse-reciprocating stoker grates, where the walls were covered with pictures of the garbage barge and lists of states ranked by different kinds of garbage production.

It also became the scene of the sorts of loud, sometimes angry debates that inevitably take place when (as someone once said of news magazine–type group journalism) many minds are brought to bear on a defenseless set of facts.

The room became known as "the Garbage Room."

The team became known as "the Garbage People."

Many life changes occurred during the six months they worked together. One reporter left *Newsday* for a job on the West Coast but said the move was unrelated to the project. Two reporters became engaged, but not to each other. And one reporter became a father.

The reporters found a throwaway society suffocating in its own trash and trapped without easy solutions. In desperation, more and more communities are turning to an answer that may be the riskiest yet: a new generation of garbage-burning plants.

Here is a summary of *Newsday*'s findings:

- The United States is spending more than $17 billion on incineration as a solution to the garbage problem. It is a national investment that could mean trading one kind of pollution for another while burdening taxpayers with enormous costs. Officials are buying incinerators in a crisis atmosphere and in ways that close off such alternatives as recycling.
- Americans are running out of space to bury the 230 million tons of garbage they produce every year. Thousands of landfills are closing, and in many areas new sites cannot be found because of environmental concerns and local opposition.
- The shortage of dump space in the Northeast has given birth to a new, unregulated industry that trucks 10 million tons of garbage a year to far-flung landfills, spreading pollution into the Midwest, raising health concerns, and costing taxpayers $1 billion a year.
- After nearly 20 years of cost overruns and plant failures, the incinerator industry is promoting a costly technology that many experts fear could be an equally risky gamble. The industry has little experience with the technology it is importing from Europe and has yet to demonstrate a long-term record of success.

- Builders of garbage-to-energy plants have downplayed and in some cases ignored the health risks that can result from fumes and ash. While debate continues over the dangers of ash, state officials allowed 30,000 tons of the waste to be spread under newly surfaced roads and parking lots on Long Island.
- Consultants with a vested interest in incineration have played a key role in shaping the nation's garbage policy—and have shared in the industry's profits. Many are former government officials who have gone to work for the industry. In the New York City and Long Island region alone, consultants and lawyers have reaped $46 million in fees.
- As the nation's garbage problem grew to crisis proportions, federal and state governments left municipal officials alone to deal with the environmental and technological challenges. Their inaction stifled efforts to encourage recycling and to cut the amount of waste Americans generate. At the same time, the federal government granted generous tax benefits to backers and operators of incinerators and required utilities to buy power from them.
- Hempstead Town has awarded garbage contracts worth more than $1 billion to a politically connected company that has convictions for bid-rigging and price-fixing and has never built an incinerator.
- New York City is trying to cope with a record increase in its daily garbage heap and a wave of illegal dumping. The city is moving to erect five major garbage-burning plants at a cost of at least $2 billion, but some experts question whether it is a wise investment.
- Recycling programs have generally failed to capture the government's attention and the nation's commitment. State and local officials are investing hundreds of millions of dollars in building incinerators and a fraction of that in encouraging recycling.

The result of the reporters' work is found in the 10-part, 55,000-word series that is being reprinted here. In the old who-what-where-when-and-why school of journalism, the "why" came last and sometimes was not dealt with at all. The larger and more complicated the story, the less likely it was to be answered. For many weeks, *Newsday* and other papers tracked the who, the what, the where, and the when of the garbage barge story.

This is our effort to explain the "why."

PART I

GARBAGE WORLD

CHAPTER 1

HIGH-STAKE RISK ON INCINERATORS

By Richard C. Firstman

Across America, they rise mysteriously from the edge of town: mighty, windowless structures of concrete and steel with soaring smokestacks and furnaces that can melt a plastic soda bottle in a tenth of a second. In municipal America's war on trash, they are the strategic weapon of choice: garbage-eating, energy-producing incineration plants.

In a country that produces more garbage than any other on the planet, a nation of dwindling environmental options, they represent one of the biggest collaborations of public works and private industry in American history. Hundreds of "waste-to-energy" plants are being operated, constructed, or planned at a cost of more than $17 billion.

But a six-month investigation has found that officials are being pressured into a solution that may be a massive environmental and economic gamble, a national experiment that could mean trading one kind of pollution for another while burdening taxpayers with enormous long-term costs. The inquiry found that the nation's garbage-incineration industry has been plagued by

3

mechanical failures that have already closed $720 million worth of incinerators and caused unscheduled shutdowns at more than half the plants now operating nationwide.

The industry is trying to recoup by selling a European technology that has virtually no operating history in this country and may not work over the long haul because of critical differences between American and European garbage.

And America is buying what the industry is selling. In the Long Island–New York City region alone, nine incinerators are due to go into operation by 1992, at a total cost of more than $2 billion.

But in many ways, the industry has not yet shown it can live up to its promise of clean, efficient waste disposal. *Newsday*'s examination shows a consistent gap between the rhetoric of its promoters and the reality of its performance.

The industry's advocates cultivate an image of smoothly functioning machines that gobble up garbage and churn out electricity. They promise an alternative to overflowing dumps that threaten to poison water supplies.

The reality is different:

The new incinerators are expensive and often unreliable.

They are being subsidized by utility ratepayers whose garbage may never go to the plants.

They contribute to air pollution and create huge quantities of toxic ash—without eliminating the need for landfills.

They undermine such cleaner and cheaper approaches as recycling. In some cases, they make it impossible.

And while the industry promotes the technology as proven, most companies selling plants in the United States have little experience building and operating them. Four of the industry's ten leading firms have never built an incineration plant of any kind; two others have built only one each.

Incinerators play a major role in garbage disposal around the world, and most experts are convinced that they are part of the solution to America's trash problem. But, in many cases, they are being sold as the only solution. They are being built with little regard for financial consequences. And they are being opened as environmental questions remain unresolved.

With garbage piling up, city councils and town boards—forced to make crucial decisions as their landfills run out of space—are getting virtually no help from higher levels of government. *Newsday*'s study found that, in the Reagan era of passive regulation, the federal government has neglected serious questions of pollution—questions about the toxins in the invisible fumes that fly out of incinerator smokestacks and in the gritty, gray ash that comes out the plants' back ends.

The momentum of the incineration movement alarms the industry's most vocal adversaries. "My worst fear is that it will be a colossal public health and financial disaster, a mistake perhaps unparalleled in recent decades," says Walter Hang, who leads an environmental arm of the New York Public Interest Research Group. "We will create a whole new class of toxic dumps for the ash and we will exacerbate the air quality that federal authorities have already said is a health hazard on Long Island and New York City."

The EPA's inaction has been part of a larger abdication by federal and state governments in dealing with the garbage crisis.

At a time when local governments were making crucial decisions about incinerators that would serve into the next century—for many officials, the most important decisions they would ever make in government—the EPA's solid-waste-policy staff was cut from 128 employees to one.

This failure to lead has left a vacuum that is quickly being filled by private industry: incinerator manufacturers and their allies—consultants, investment bankers, lawyers, and influential former government officials who have shared in the lucrative business of resource recovery. In the region, municipalities have paid these advisers more than $46 million.

Incinerator operators contend that, like any new technology, the industry has suffered through an inevitable period of trial and error. They dismiss the suggestion that the plants cannot work in this country.

"We're offering mature plants," says David Sussman, environmental director of Ogden Martin Systems, a large incinerator manufacturer. "It's like driving home a BMW—when you get home you know it's going to work. I've heard all these arguments: the garbage is different, the operators are different, the environment is different. But none of it is true. What you've got out there is the future."

The future comes at a time when Northeastern garbage is being trucked to the American heartland as landfills across the country close at a rate of ten a week.

So much garbage has been shipped off Long Island that, on average, a tractor-trailer carrying 40,000 pounds of trash would enter the Long Island Expressway every six and a half minutes. Garbage has been the Island's leading export by weight—surpassing ducks, clams, and potatoes.

The future was heralded in spring, 1987, when a barge full of Long Island garbage sailed the Atlantic Coast in search of a friendly port. Four states and three foreign countries refused it entry.

As the barge floated in New York Harbor, and lawyers and politicians debated its fate in court, *Newsday* began looking into the crisis it had come to symbolize.

Over the next six months, reporters reviewed tens of thousands of pages of documents, interviewed more than 500 people, toured garbage plants across the country, and surveyed operators of 227 incineration plants and environmental officials in 55 states, 5 territories, and the District of Columbia. What they found was a mundane municipal enterprise that had taken on a life of its own.

"We are in the midst of a garbage revolution," says David Gatton, director of policy analysis for the U.S. Conference of Mayors' resource recovery association. "For the first time, this country is realizing that garbage is coming above ground. If we don't implement solutions quickly, we will definitely have a national crisis."

If it is a crisis, it is one that has been building for 50 years—the product of a disposable, over-packaged society. It is a society that has never dealt with its growing production of garbage, never embraced the idea of recycling, and now finds itself starting to suffocate in its own trash.

Two solutions have already failed. First, the early incinerators polluted the air. Then, landfills overflowed and threatened groundwater. In desperation, communities have begun shipping their garbage hundreds of miles—only to find that long-distance trucking is expensive and exports pollution as well as garbage.

As the circle closes, the country now turns to the latest answer: a solution that may create as many problems as it purports to solve.

Welcome to Garbage World.

For your plastic bag of garbage, it is a world of infinite possibilities. It may be trucked to your neighborhood landfill, to be buried perilously close to the source of your drinking water. It may be carted to Ohio or Pennsylvania in the same trailers that will return east with crates of Midwestern produce. It may be shipped to Staten Island, where the garbage mountains of Fresh Kills are growing at a rate that, by the turn of the century, will make the largest town dump in the world one of the highest points on the Eastern Seaboard. Or, your plastic bag of garbage may see the world, piled onto a barge to nowhere.

Garbage World is the municipal equivalent of the federal deficit: How long can it continue, how far can it go? Across America, nervous politicians are caught between conflicting commercial and political interests, and between converging realities: landfills posing environmental hazards as they creep closer to capacity, and a throwaway society that offers no indications that it will suddenly begin throwing away less.

In this atmosphere of crisis, a kind of municipal-industrial

A couple that for a while had their names in the news as often as film stars—the barge Mobro 4000 *and its tug, the* Break of Dawn.

Newsday/Dick Kraus

complex is in the midst of a mad, perhaps misguided, dash for salvation. It is called resource recovery—a euphemism that implies environmental virtue but that tends to obscure persistent questions of safety, competence, economics, and the environment itself.

On Long Island and in New York City, the concentration of nine plants due to open by 1992 and four others in the planning stages—nearly one for every town in Nassau and Suffolk and every borough of the city—raises questions about environmental effects.

They are the new incinerators: sleek concrete structures that could be I.M. Pei works with smokestacks. They seem logical— hundreds of tons of garbage burned every day, turned almost magically into kilowatts of energy. These new "mass-burn" plants, in which truckloads of unsorted garbage are fed into furnaces that boil water that produces steam that turns turbines that produce electricity, would seem to be the perfect solution to the puzzle of American garbology. And municipal America likes the idea: By 1992, there are expected to be more than 200 garbage-burning plants in this country.

But it is an idea advanced by an industry that often has been better at selling incinerators than at operating them. Although the technology is fallible, the operators inexperienced, and the result potentially dangerous, garbage-to-energy plants are sprouting across the American landscape like so many 7-Elevens. There are now more than 100 plants operating in the United States, with another 115 under construction or development and scores of others being discussed—a national investment, thus far, of more than $17 billion, a figure that could nearly double by the turn of the century.

It is not hard to understand why municipalities are practically lining up to buy incinerators. Nearly 3,000 municipal landfills have closed since 1982 because they were either filled or environmentally dangerous. And half the remaining 8,736 landfills are expected to run out of room by 1997. As the American way of garbage approaches a critical mass, local officials feel pressured to do something. While incinerators may represent a trade-off, it is a gamble many seem willing to take.

"In Huntington we had no more space for landfilling. Same in Islip and in Nassau County," said Huntington Supervisor John

O'Neil, whose town has signed a contract for a new incineration plant. "How far into the future can we do this?"

"I think it will work," said former Babylon Supervisor Anthony Noto, who has more recently gone into the resource recovery consulting business.

In many cases, Long Island officials have come into office with few choices. Some options have been preempted by the failure of their predecessors, or by state legislative mandates to close land-fills. Still, critics claim, officials are ignoring such combined approaches as scaled-down incinerators and stepped-up recycling programs.

Although it sounds straightforward, modern garbage-burning machinery—with its elaborate arrangement of electrostatic precipitators and reverse-reciprocating stoker grates and polychlorinated dibenzofuran-trapping bag-houses—is a tricky technical enterprise. It is somewhere between an old-fashioned town dump and a modern nuclear reactor. Experience is important, and in many cases it is missing.

One of America's largest garbage-burner manufacturers, Ogden Martin Systems, which is under contract to build 14 plants at a cost of $2.3 billion, had only three plants in operation by winter, 1987. None was more than 18 months old.

In a report issued in the fall of 1987, Moody's Investors Service said that European-style incineration "is not yet a 'proven' technology" in the United States, that unscheduled shutdowns for all types of plants appear to be increasing, and that "such a project clearly entails major risks."

"We're still in the infancy stage with this industry," says Edward R. Kerman, a vice president of Moody's. "There are only a handful of [mass-burn] plants nationally with any kind of operating history."

Still, cities and towns across the country are rushing to judgment. A striking example of the gamble is in the town of Hempstead, home of one of the industry's worst disasters. Seven years after the failure of its first resource recovery plant—an endlessly controversial $135 million machine that operated on and off for only 18 months before fears about pollution put it out of business—Hempstead hired an inexperienced firm to build a second plant on the same site.

The firm, American REF-FUEL, has never built or operated a

resource recovery plant. It offered to build one of the nation's most expensive plants, with a price tag of $370 million. Its proposal called for Hempstead to take on risks normally assumed by the builder. But American REF-FUEL won the job—after Hempstead officials took only four days to review hundreds of pages of bid documents. The company's parent, Browning-Ferris Industries, also made a move that has become common in public-works industries: It hired a politically connected lawyer. In this case, it was Armand D'Amato, then a state assemblyman, whose brother is U.S. Senator Alfonse D'Amato, the town's supervisor when the first Hempstead plant was built.

At sunrise one day in the spring of 1987, before a thousand cheering spectators, the guts of the first Hempstead garbage plant collapsed under the force of 72 pounds of dynamite. For the next generation, only a concrete floor, a few offices, and a metal facade were left to stand.

Garbage World has always been a little strange. In 1899, when rural communities were feeding garbage to hogs, Manhattan was dumping its trash into the ocean. That year, the New York Sanitary Utilization Company opened one of the first garbage incinerators in the United States. Five years later, the incinerator was producing the electricity to light the Williamsburg Bridge.

In 1931, the U.S. Supreme Court banned ocean dumping. Landfills were filling up. And incinerators were rising across the cityscapes of America. Then, as now, they were seen as the Big Solution. This is what New York Mayor Jimmy Walker had to say that year: "We have found that everybody is in favor of incinerators, but nobody desires them as neighbors. I am assured, however, that the plants will be scientifically built, and their operation will be attended neither by objectionable odor, noise, nor smoke. We have to build these plants to dispose of our refuse. We cannot kick the garbage around the city."

And so, the smokestacks went up. But by the 1970s, stricter air-quality standards were forcing them down. And back the garbage went, into big, sandy holes in the ground. This time, garbage engineers designed clay and plastic liners and called the dumps sanitary landfills.

But with a shortage of landfill space, people still felt the need to burn. During the 1970s, as oil prices were escalating, the early generation of resource recovery plants was being promoted more as energy suppliers than as garbage burners. Most of them, including the Hempstead plant, were known as refuse-derived fuel plants, a type that requires sorting of the garbage for metals and other incombustibles and produces a papery fuel that is used to create energy. But they were plagued by chronic mechanical failures, explosions, and pollution problems.

The '80s-generation of resource recovery does not bother with sorting trash. Everything, except for the occasional Volkswagen fender, goes into the fires—a method that discourages recycling. Incinerator operators want guarantees that they will get enough garbage—especially plastic and paper, which burn nicely and provide a lot of energy—to keep their plants running. Still, the latest technology has not yet proven to be much more reliable than its predecessor. In a nationwide survey, *Newsday* found that many incinerators are routinely shut down because of mechanical problems.

At the heart of the problem is that serious obstacles stand in the way of transplanting European incineration technology to this country. Because it contains more plastics and includes products that Europeans recycle, American trash tends to produce acid gases that add to pollution problems and corrode plant equipment, requiring time-consuming repairs.

Economics is also a factor. In this country, more than in Europe, resource recovery is a money-making enterprise. Energy sales account for more than half the revenues of most large plants, and their operators are under great pressure from investors to turn profits. And so, American plants are generally bigger and their steam-generating boilers run hotter than their European counterparts, in order to sell more electricity.

In many cases, such upscaled plants have courted financial disaster. A plant that was opened in Saugus, Massachusetts, in 1975 is considered a success because it is still operating. But during one four-year period, the owners needed a federal bailout to pay for $11 million worth of repairs. A plant in Florida, built in 1983, had so many problems when it opened that it was unable to meet its bond payments for nearly two years.

Although most incinerators are produced by the private sector, their costs are generally underwritten more by the public than by the firms that build them. Under the complex financial structure of resource recovery, residents pay the multimillion-dollar costs of plants in three ways.

Garbage collection fees and taxes—charged either by private carters or municipalities—must rise to cover the cost of higher "tipping fees" at the incinerator's gates. Long Island planning officials say incineration may mean a tripling of the $230 a year the average family now pays for garbage disposal.

Exemptions from federal taxes for bond holders mean the construction of plants is being subsidized by federal income-tax payers. And electric ratepayers pay roughly half the operating costs of a new plant, because a federal law requires local utilities to buy the power that the plants produce—at a rate that utilities say is nearly twice what it would cost the utility to produce the power itself.

All of the Long Island Lighting Co.'s (LILCO's) customers must subsidize the local garbage plants, even though only some of them will have their garbage burned by them.

"There are two calamities occurring—one environmental, one economic," says Neil Seldman, head of the Institute for Local Self Reliance, based in Washington, D.C. "More lead will be put into the air . . . and it's going to break the bank."

Even if a plant runs efficiently, there are serious questions about its impact on the health of its neighbors. Scientists are still debating whether ash is hazardous, whether dioxins cause cancer, and whether a variety of other substances that come out of resource recovery plants threaten human health. A report released in 1987 by the federal Environmental Protection Agency found that emissions from incinerators do pose threats to the people exposed to them. The EPA study estimated that, if all planned incinerators are built, anywhere from 4 to 60 people nationwide might develop cancer each year because of the emissions.

While the risk seems relatively small, critics contend that the EPA's study underestimates the potential peril.

Dioxins, for instance, are organic compounds formed during the burning process. At high doses, they kill hamsters in the laboratory, and scientists suspect that they promote cancers in people. Similarly, the combustion process releases low levels of acid gases and toxic metals, particularly lead, that are allowed to escape into the atmosphere at varying degrees.

Pollution testing and control has been a haphazard area of resource recovery—almost an afterthought. Generally, state and federal officials have allowed the industry to build first and answer questions later, if at all. In the absence of any federal regulations to reduce levels of emissions, states have imposed widely varying limits.

Pollution-control equipment, meanwhile, differs from plant to plant. While an incinerator in Oregon boasts a sophisticated system of devices to remove toxic particles and gases from smokestack emissions, there were, in 1987, 16 operating plants across the country that had no control equipment at all, and 18 more were being planned.

In Florida, officials have never measured dioxin emissions at the state's nine resource recovery plants. They say the cost— $100,000 a test—is too high. In Massachusetts, where resource recovery incinerators have operated for more than a decade, it was not until 1986 that state regulators performed their first such test. They found so much dioxin coming out of one plant that they shut it down—but only after it had been operating for more than two years.

Independent, credible information is hard to come by, because tests are often conducted by private plant operators whose results are difficult to compare. Estimates of pollution from plants that are only in the planning stage are even more unreliable. Bernd Franke, a consultant commissioned by *Newsday*, found that estimates of pollution that might come from six of the Long Island– New York region's 13 planned plants vary widely, despite the fact that the plants will have similar technology and pollution-control devices.

Not even plants such as the one in Marion County, Oregon, and another in Baltimore, promoted by the industry as two of the best, have unblemished environmental records. When the Oregon plant could not meet the state's emission limits for nitrogen oxides, the state raised the limits so it could. At Baltimore, no tests have ever been performed to determine emissions of dioxins.

Even when airborne pollutants are reduced, there is another Garbage World dilemma: The toxic particles have to go somewhere, and they wind up in the ash. In a report released in November, 1987, the EPA said that incinerator ash contains dangerous levels of cadmium, lead, and dioxin. Despite the tests, the EPA says it is planning to recommend that ash not be treated as a hazardous waste—a decision likely to be applauded by the incinerator industry, which does not want to pay the high cost of disposing of ash as a toxic material.

The issue has already come up in Glen Cove, whose incinerator was found in January, 1987, to be producing toxic ash. As state officials were arguing the question of where to put an ash disposal site—a debate that may be repeated across the country in coming years—they allowed a contractor to use 30,000 tons of ash to build roads and parking lots all over Long Island.

"There's no such thing as a 'safe' incinerator because there's a Catch-22," contends Paul Connett, a researcher and dioxin expert at St. Lawrence University. "The better the incinerator is at protecting the air, the more toxic the ash is going to get."

The industry maintains that dioxins are not dangerous at low levels and that state-of-the-art controls eliminate other potential pollution problems.

Says Sussman of Ogden Martin: "The risk from emissions is far less than from many other combustion sources that we deal with every day—automobiles, fireplaces, power plants. Those are the biggies. Your risk of being murdered in your lifetime is one in 150. . . . There aren't any environmental risks that are even close to that."

More objective analysts are a rare breed in Garbage World. "If I were to look at the available data as a report card, I'd have to give municipal waste combustors an incomplete," says one, John Skinner, director of the EPA's Office of Environmental Engineering and Technology Demonstration. "We need more data and we need more homework."

That kind of research has not flowed freely from Skinner's agency. Only recently has the EPA—under pressure from Congress and environmental groups, one of which sued the agency—begun developing specific, nationwide emissions limits. But the standards will not be in force until 1990—when, critics contend, they may be too little, too late. By then, scores of plants already will be operating.

Relentless fears about pollution and a contempt for anything to do with solid-waste disposal have brought a resurgence of the single-issue civic action group. Across the country, they adopt acronyms like RAGE and CLOSE and fight the incinerators—a conflict for the '80s, reminiscent of the anti-nuclear power movement born a decade ago.

The views on all sides are passionate. From the pro-burn establishment comes Dan Madigan, a vice president of Wheelabrator Environmental Systems who oversees the company's plant in Saugus, Massachusetts. "Just because this plant has a chimney, the so-called environmentalists will use it for their own political gain," he says during a tour of the plant. "No matter how well you do, the regulators will try to make you do better. How much better can you make it? The fact is, if you light a match you're going to have an emission. I think we should take a match to the town dump every Saturday. I think we should go back to burning leaves."

From the resistance comes Walter Hang, a kind of guerrilla leader of the anti-incineration movement: "It will mean we will be locked in for decades to this antiquated garbage-burning technique, and there will be no way to turn back. It will be one of the biggest failures in environmental history."

They are all characters in Garbage World—the government-bureaucrats-turned-industry-advocates who offer to eat spoonfuls of ash; the holistic-disposal gurus who seek deliverance in piles of compost; the environmental entrepreneurs who see garbage as gold; the incineration evangelists and defenders of the natural universe who duel in the courtyard of public opinion.

Flat opposition to incineration tends to ignore one crucial fact: The garbage has to go somewhere. And landfills are not a remedy. They produce plumes of pollutants—methane gas and a witch's brew called leachate that can poison drinking water and foul waterways. But remaining unresolved is the question of whether dumps are more dangerous than incinerators.

One school of thought promotes a combination approach: a little dumping, a little burning, and more recycling. That path has divided the environmental-watchdog community into two camps: those who support such a compromise approach and those who hold out for intensive recycling and other methods of garbage reduction.

While state governments support recycling in theory, they have

generally failed to take up the cause aggressively—partly because, in many places, incineration and large-scale recycling are
fundamentally incompatible. Incinerators need primary recyclables like paper and plastic to operate at a profit.

In New York State, for instance, officials have announced that
they want to reduce the state's garbage production by ten percent
by 1997, and recycle 40 percent—an eightfold increase. But such
ambitious goals of garbage rehabilitation seem all but forgotten
in the rush to burn.

"Everyone agrees that landfilling is bad for the environment,
but many municipalities feel that one big plant is going to solve
their garbage problems," says Howard Levinson of the U.S. Office
of Technology Assessment. "They aren't looking sufficiently at
recycling efforts. Everyone seems to be going whole hog for incineration."

It is not a coincidence. Across the nation, the sanitary landfill
is slipping into obsolescence. Space is running out; anxiety about
groundwater pollution is growing. On Long Island, the situation
is particularly acute: In a place where every man, woman, and
child contributes more than a ton a year to the local waste
stream, the state has ordered most landfills closed by 1990 because of fears that the poisons of decomposing garbage will seep
through their clay and plastic liners, possibly contaminating the
main groundwater supply.

And in New York City, the nation's largest municipal trash
producer, garbage is threatening to choke the city. Nearby landfills are closing or raising their fees, forcing a record increase in
garbage at the city's dumps. Meanwhile, carters are dumping
garbage into building foundations and onto lonely city streets.

As the No-Room-At-The-Dump signs go up, the trucks get
rolling. In 1988, Northeastern cities and towns paid nearly a
billion dollars to send their garbage to landfills in distant territories—places like Pennsylvania, Kentucky, and Ohio, where the
shipments are welcomed with such homespun roadside greetings
as "Don't Kill Our Children With Out of State Trash."

This new industry, known in '80s-speak as the spot landfill
market, brings together the worlds of waste management, agriculture, and interstate cruising. After a freelance driver dumps
his unseemly freight into a Midwestern pit, he may pick up a load
of produce for the trip back east.

Men dispose of a bathtub and other trash on a Saturday at Brookhaven Town landfill.

And, occasionally, someone comes up with a new idea, such as loading garbage onto a barge and floating it to Mexico.

Against a backdrop of such environmental burlesque, resource recovery has built an aggressive growth industry—a sophisticated, big business out of a once-primitive and insular service trade dominated by mobsters. The waste-disposal business of the '80s is less about collecting garbage than about making it disappear. And in the municipal offices of America, the resource recovery industry has found a receptive audience.

With little leadership from higher levels of government, local officials find independent counsel a rare commodity. And so, much of their advice comes from the industry itself, or from consultants tied to the industry—part of a political-industrial network that wields much of the subtle power in the American garbage business.

The municipalities of New York City and Long Island have paid more than $46 million to solid waste consultants for legal advice and engineering services that tend to promote the industry. Although they are ostensibly objective analysts, many of these firms have a financial self-interest in resource recovery.

"The first thing that government typically does when it is faced with a big environmental problem is it goes to the consultants," Hang says. "The consulting engineers don't really make money by saying to a community, 'You can solve your garbage problems once and for all by drastically reducing the amount of garbage you produce.' . . . Consulting firms make money when they say, 'We'll design an incinerator, we'll help you finance it, we'll help you build it, we'll help you operate it.' "

The industry also has been a shrewd marketer of its product—packaging incinerators in handsome, angular facades and employing the latest business-school sales techniques. Some companies arrange trips to Europe so local officials can tour garbage plants there.

Resource recovery firms sometimes enhance their proposals with offers they hope communities cannot refuse. In Los Angeles, bidding competition between Ogden Martin and Signal Environmental Systems included a promise of a $10 million community "betterment fund"—with the money coming not from the company but from the $235 million bond issue sold to finance the plant. There were also campaign contributions to City Council candidates—a gesture of goodwill that has become a common part of the industry's big sell.

Business has not suffered by the close relationship of industry and government. At a time when he was serving as a founding member of the state Legislative Commission on Solid Waste Management, an advisory panel that was set up to help solve New York's garbage problem, Armand D'Amato was promoting incinerators for his other employer: Browning-Ferris Industries. Over at Signal, new employees included former New York Lieutenant Governor Alfred DelBello and William Rauch, former press secretary to and co-author with New York City Mayor Edward Koch.

And at a convention of the National Solid Waste Management Association in Boston in 1987, Ogden Martin officials David Sussman and Garrett Smith stood side-by-side in their company's hospitality suite, shaking hands with mayors, council members,

The American Way of Garbage
The national picture, based on a Newsday survey of 50 states, five territories and the District of Columbia

How much is thrown out
Americans throw out 227.1 million tons of garbage a year, enough to fill about 187 World Trade Centers, or about one of the twin towers per day

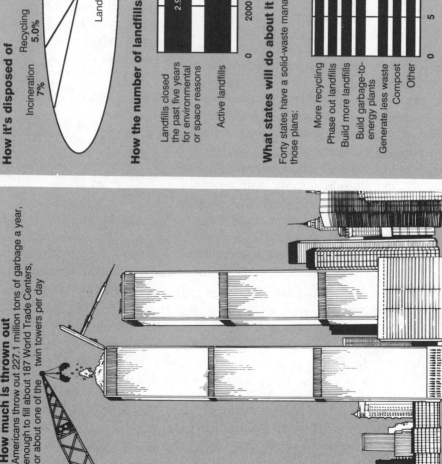

How it's disposed of

Landfills 87.1%

Incineration 7%

Recycling 5.0%

Other 0.9%

How the number of landfills has fallen

Landfills closed the past five years for environmental or space reasons	2.991
Active landfills	8.801

0   2000   4000   6000   8000   10,000

What states will do about it
Forty states have a solid-waste management plan. Here are the objectives of those plans:

More recycling	30
Phase out landfills	14
Build more landfills	16
Build garbage-to-energy plants	28
Generate less waste	25
Compost	10
Other	14

0   5   10   15   20   25   30

SOURCE: Newsday survey

Newsday/Steve Madden

A GLOSSARY OF "GARBAGE WORLD" TERMS

By Irene Virag

As with most places in this technological age, even Garbage World has a vocabulary all its own. The techno-babble describing what we throw away and how we do it can be as foreign sounding as computer manuals and legal contracts, apartment leases and appliance warranties, and instructions for putting together knockdown furniture.

The glossary below will help decipher the language of Garbage World.

Bottom ash—The residue that collects in the bottom of the burning chamber of an incinerator.

Bypass garbage—The buildup of refuse when an incinerator is shut down.

Compost—Decayed organic matter used for fertilizing and conditioning the land.

Electrostatic precipitator—An antipollution device that electrically charges particles and draws them into a collector to keep them from getting into the atmosphere.

Emissions—Substances that are discharged into the air from the stacks of a garbage-burning plant.

Fly ash—The fine particles of ash in flue gases produced by burning garbage.

Incinerator—A furnace for burning waste.

Leachate—Rainwater that washes through a landfill and picks up pollutants.

---

and sanitation commissioners and joking about eating some ash residue to prove that it is safe. Sussman and Smith used to work for the EPA.

People have been talking about garbage for a long time.

"If the entire year's refuse of New York City could be gathered together," *Scientific American* mused in 1912, "... the weight of this refuse would equal that of 90 such ships as the Titanic."

"Some day," the *Milwaukee Sentinel* sighed in 1892, "a Milwaukee Moses may lead us to the promised land, where there is neither garbage nor complaining about garbage."

Sussman has a theory about why it strikes such an emotional chord in people. "I'm a Freudian," he says. "The earliest discomforts you experience, and the first thing you're proud of as a baby,

Mass burn—An incineration system in which nothing is sorted. Everything goes into the same fire. Steam or electricity is normally produced.

Municipal solid waste—Garbage, including household waste, street litter, commercial refuse, abandoned automobiles, and ashes.

Noncombustible wastes—Metals, tin cans, foil, dirt, gravel, bricks, ceramics, glass crockery, ashes.

Refuse-derived fuel—An incineration system that requires sorting of garbage for metals and other noncombustibles and produces a fuel used to create energy.

Resource recovery—The extraction of materials or energy from wastes.

Scrubber—A device for removing pollutant dust particles from the air stream of an incinerator.

Source separation—Sorting garbage at the point of generation and putting specific discarded materials such as newspapers, glass, and metal cans into containers for separate collection.

Tipping fee—The charge to unload waste materials at an incinerator or landfill.

Waste stream—All the garbage produced by the community that is disposed of in incinerators and landfills.

is a waste management function. We are conditioned to not want to be around anything to do with waste."

But it is all around, and getting worse. Americans produce about 230 million tons of refuse a year—5.1 pounds per person a day. That is more garbage than any other country in the world, even more than China, which has four times the population. New York–area residents are among the most productive in the nation. Long Islanders contribute 6.8 pounds a day per person; city residents produce 5.5 pounds.

By volume and content, American garbage reveals a disposable society that tends to prefer convenience to conservation, short-term gratification over long-range resourcefulness.

"We think of creating a Styrofoam container to hold a hamburger for five minutes," says A. James Barnes, deputy administrator of the EPA. "Then it may be thrown into the environment in a form that could last for several thousand years."

Or it may be burned instantly in a resource recovery plant. But while the plastic foam may disappear, the larger problem will not. That is the ultimate, inevitable reality of garbage. It does not go away. The early incinerators were offered as a solution to the piles of garbage of an industrialized nation. Landfills were presented as an answer to the pollution of the incinerators. And now, resource recovery plants have been offered as a solution to the landfill problem.

But in the circuitous reality of Garbage World, they may be the next big problem.

CHAPTER 2

THE ALL-CONSUMING LIFESTYLE

BY IRENE VIRAG

It happened, as it always does, without anyone in the Sigmann family even noticing.

At six in the morning, Deneen Sigmann was half asleep in her four-bedroom ranch on Long Island when she opened her first pack of Salem Lights and threw the cellophane wrapper into the trash. It floated as silently as an autumn leaf into the 30-gallon plastic bag that lines the brown Rubbermaid garbage can in the kitchen. The wrapper joined the previous day's newspaper, a used Brillo pad, a mound of leftover ziti, an empty shampoo bottle, a crumpled bag of potato chips, a broken glass, a banana peel, a plastic yogurt container, four packets of Sweet 'N Low, and a half-full can of beef stew for dogs.

Deneen Sigmann lit a cigarette and headed into the bathroom. As she left the kitchen, her children were fixing their own breakfasts. Twelve-year-old Jennifer poured the last of the Cheerios into a bowl and threw the box away. Sixteen-year-old Tom tore open two individually wrapped store-bought fudge brownies. He ate the cakes, wiped his mouth with a paper towel, and tossed a wad of paper packaging into the Hefty Cinch-Sak. Eight-year-old Cathy finished the Count Chocula cereal and the milk—then shoved the cardboard box and the one-gallon plastic jug into the bulging garbage bag.

When Deneen Sigmann returned to the kitchen, the garbage can was close to overflowing. Waltie, Deneen's 21-year-old son, had made a brown-bag lunch—four slices of stale white bread and the plastic wrap from the bologna were added to the can. Deneen drank her fourth cup of black coffee, then dumped the grounds. She emptied an ashtray—a cigarette butt stuck in a glob of spaghetti sauce that had been discarded the night before, and the ashes from two packs of Salems descended like a fine snow that hides the litter underneath.

Before leaving for work, Walter Sigmann, Deneen's husband, stuck his hand into the bag and shoved the contents deeper into the Rubbermaid can. He crushed cereal boxes and milk containers. He squished the pasta into the cherry-vanilla yogurt. He made room for more garbage.

Later that weekday evening, sometime between the after-dinner cleanup and bedtime, Walter Sigmann pulled the 30-gallon Hefty Cinch-Sak out of the Rubbermaid trash can. He carried it outside and plopped it down in front of the garage next to three equally stuffed bags, one of which had been ripped open by a hungry dog or maybe a raccoon. An onion and an apple core lay on the driveway.

By Sunday, six bags blocked the garage door. Walter Sigmann piled the week's worth of garbage into his van. He drove five miles to the Brookhaven Town landfill and waited on line 45 minutes to pay a $2 entry fee. He added his bags to the vast wasteland.

When Walter Sigmann returned home, a clean 30-gallon plastic bag lined the trash can in the kitchen.

Deneen sat at the table smoking a cigarette, cutting quilted fabric to fashion into Christmas stockings. Walter ate a piece of store-bought peach pie and clipped coupons. A few minutes later, Deneen emptied another ashtray and Walter got rid of the Sunday newspaper.

It was happening again, as it always does, without anyone in the Sigmann family noticing.

The way it happens in most American homes, where the disposal of personal trash is as much a part of daily life as eating and drinking. It is a matter of habit. And habits can be hard to break. When it comes to garbage, America's routines are rooted in the psychology of a disposable society that seems as if it were designed to keep us from solving the mess. A society dedicated to plastic and paper and prepared foods, to the carryout and the convenient, to disposable diapers and dishes, throwaway razors and roasting pans. A society that gorges on fast-food burgers served in Styrofoam—a nonbiodegradable petrochemical plastic product that stays forever in landfills or produces toxic fumes when burned.

A culture of consumption reflected in statistics—the United States produces almost 230 million tons of garbage a year. It has twice as many people as Japan but produces nearly four times as much trash.

A culture of planned obsolescence in which very little is built to last a lifetime and it is more practical to add broken clocks and calculators and can openers to the nation's landfills than to repair them.

"We see what's around us today, everything we have, and we believe it's God-given," says William Rathje, an anthropologist at the University of Arizona who has studied tons of American garbage during the past 14 years. "We believe that somehow God or IBM or somebody is going to come up with a solution and we're not going to have to change our behavior."

Rathje's ongoing study of Tucson's trash—known simply as the Garbage Project—reveals that waste is imprinted in the national psyche. "The first bag I ever opened had a whole T-bone steak in it, fully cooked and wrapped in a paper towel," he recalls.

One of Rathje's findings would anger starving people around the world—Americans waste 15 percent of all solid food they buy. The project also showed that middle-income families throw away more food than their upper- or lower-income counterparts, that wealthy people buy inexpensive foods for themselves but expensive foods for their pets, that a bag of garbage loaded with frozen food packaging is likely to be just as stuffed with rotten vegetables and spoiled fruit, and that the length of the stem cut off fresh asparagus increases with income levels. The study also documents the country's growing consumption of plastic; between

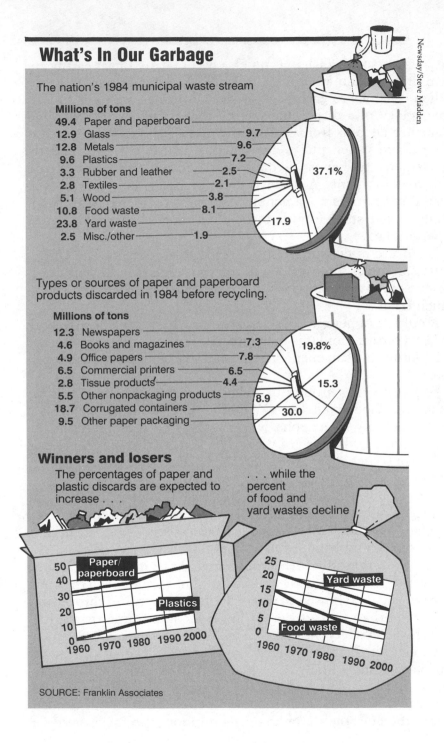

What's In Our Garbage

The nation's 1984 municipal waste stream

Millions of tons

49.4	Paper and paperboard	37.1%
12.9	Glass	9.7
12.8	Metals	9.6
9.6	Plastics	7.2
3.3	Rubber and leather	2.5
2.8	Textiles	2.1
5.1	Wood	3.8
10.8	Food waste	8.1
23.8	Yard waste	17.9
2.5	Misc./other	1.9

Types or sources of paper and paperboard products discarded in 1984 before recycling.

Millions of tons

12.3	Newspapers	19.8%
4.6	Books and magazines	7.3
4.9	Office papers	7.8
6.5	Commercial printers	6.5
2.8	Tissue products	4.4
5.5	Other nonpackaging products	8.9
18.7	Corrugated containers	30.0
9.5	Other paper packaging	15.3

Winners and losers

The percentages of paper and plastic discards are expected to increase . . .

. . . while the percent of food and yard wastes decline

Paper/paperboard

Plastics

50 40 30 20 10 0
1960 1970 1980 1990 2000

25 20 15 10 5 0
1960 1970 1980 1990 2000

Yard waste

Food waste

SOURCE: Franklin Associates

1980 and 1985, there was a 40 percent increase in castoff plastics.

On a more personal scale, the Sigmanns' Cinch-Sak reflects a similar culture. Walter and Deneen both work—he is a delivery person for a wholesale stationery company; she is a billing clerk at Brookhaven Memorial Hospital—and their garbage reflects their busy schedules. It is littered with the cardboard cartons from frozen fish fillets and instant oatmeal, mushroom soup and macaroni and cheese dinners, with stale bagels and cinnamon buns, cooked pasta and moldy cheese, raw chicken cutlets and seasoned bread crumbs, half a head of lettuce and liverwurst.

All these things find their way into the family's garbage can in the course of an ordinary day, but like most Americans, Deneen and Walter Sigmann usually do not notice the bulging plastic bag until it is overflowing. And even then, they do not consider its contents. The advertising slogan for Hefty Cinch-Sak echoes the collective state of unconcern: "Never touch garbage again." Wrap it up and give it away. Out of sight, out of mind.

The commercial catch phrase is perhaps a new national credo, the American way of garbage as described by Deneen Sigmann: "I don't want to see the stuff or touch it or think about it. I don't have time to worry about garbage—just get it out of here."

It has been this way for centuries.

Long before plastic liners and rubber cans, the streets were America's dump. When New York City was still New Amsterdam, the government passed a law prohibiting citizens from dumping rubbish, filth, and dead animals into the streets. But later, colonists still threw their garbage out the door for the pigs to eat. In the 1800s, cows and dogs roamed the dusty streets, and dead horses were as common as kerosene lamps. A century before the Islip garbage barge put to sea in search of a home, street cleaners in New York City loaded refuse on scows, dumped it at the mouth of the harbor, and hoped the tide would take it away.

Garbage was the reality that belied the romantic notions of the past. Nobody wrote songs about the trash all around the town. Mamie O'Rourke and her boyfriend would have stumbled over horse dung and fish heads if they tried to trip the light fantastic on the sidewalks of the real Old New York.

East Side as well as West Side, the thoroughfares were piled so high with rubbish that the editor of the *Evening Post*, in the spring of 1831, compared the filth in the main avenues to the Alps and the Andes. A few years later, the *Daily Times* decried the dead rats and garbage "undergoing a process of fermentation in a pool of stagnant putrid liquid" in the gutters of Grand Street.

In 1837, the metropolis spent $1,200 to remove 347 dead horses, 1,182 dogs, 3,091 cats, and 9 cows from the roads. Not only were manure piles a major problem, but the Deneen Sigmanns of the past—without galvanized garbage cans and plastic bags—dumped potato parings and other kitchen slop into the street. Privy tubs and cesspools overflowed into the gutters, creating clouds of flies and a sickening stench.

Although itinerant ragpickers called chiffoniers scavenged through the refuse, the most visible sanitation crews were four-footed—loose pigs cleaned up the garbage. But they also created their own. The city was caught in a vicious cycle—as long as garbage was tossed in the streets, the hogs flourished, and as long as hogs roamed the streets, it was much simpler to throw garbage into the gutters. Newspapers railed against the roaming legions of 10,000 hogs. As always, change came slowly. Cleanups were inspired only by crisis—the threat of yellow fever or cholera, an upcoming election—and the zeal to sweep up the streets dissipated as soon as the epidemic ended or the campaign was won.

The 19th-century pig problem illustrated a constant conundrum of garbage disposal, one demonstrated today by the controversy surrounding landfills and incinerators: Solutions bring their own problems. Eventually, thousands of pigs were rounded up and destroyed. But the hog cleanup was a blow to the poor, who could afford to keep pigs only because the streets were paved with feed. There was more garbage in the public view. And the resulting increase in offal caused the contractor who collected it to demand a pay raise from $9,000 to $12,000 a year.

It was not until the latter half of the century that most of the pigs had been removed from the streets and human crews took over. More sophisticated steps were still to come—the first U.S. garbage incinerator on Governors Island in 1885, and even a short-lived attempt at recycling near the end of the century that saw New Yorkers putting rubbish, food wastes, and ash in separate receptacles.

Whether it was burned, separated, eaten by pigs, or taken to sea, the garbage kept piling up. In Manhattan at the turn of the century, an average of 612 tons of garbage was collected daily. Each American contributed 300–1,200 pounds of ash, 100–180 pounds of food waste, and 50–100 pounds of rubbish yearly—a total of less than 1,500 pounds. Today, each American creates about 1,900 pounds of garbage a year.

Then, as now, the mass of trash did not go completely unnoticed. In 1907, a conservationist named Austin Bierbower, writing in a progressive journal, got to the bottom of the heap with an assessment that holds true today: "Nowhere in the world is there such a waste of material as in this country. . . . we destroy perhaps as much as we use. Americans have not learned to save, and their wastefulness imperils their future. Our resources are fast giving out. . . . In passing the alleys of an American city, a foreigner marvels at the quantity of produce in the garbage boxes."

The curbside cans of today offer equal cause for wonder. If garbage fills our lives, our lives also fill our garbage. The act of throwing something away is so unconscious that most people do not even know what is in their garbage. Unlike the Japanese, who have to sort their refuse carefully, few Americans count how many paper towels are tossed into the trash or worry about where the Q-Tips go.

The result is that all the things we are can be found in our garbage. Garbage reveals the self-delusions of ordinary people, it lays bare the good intentions never carried out. It is something of a repository for truth. "People can lie," Rathje says. "Garbage doesn't lie. . . . People will tell you what they do or think they do, or what they want you to think they do. Garbage is the quantifiable result of what they actually did."

Which is why police officers and private investigators, market researchers and reporters, have been known to go through the trash of people they want to find out about. Secrets end up in the garbage—secrets people might not tell their best friends or census-takers, secrets they might not admit even to themselves.

"We don't see garbage as the remnants of our behavior," Rathje explains. "We see it as yucky disgusting stuff that has nothing to do with who we are. But garbage is a vital, important, relevant part of our lives."

A bag of garbage provides a picture of our personal lifestyles— and it also fits the individual into a societal frame. Rathje found that garbage charts the changes in society: As more women joined the work force, his researchers noticed an increase in takeout and frozen-food containers; as the carry-your-own music craze took hold of America, they observed more AA-batteries, the size used to power Sony Walkman radios. Two of the biggest banes for American landfills reflect two of the biggest boons for society—old phone books, the symbols of our almost universal dependence on the telephone, and worn-out tires, the remnants of our love affair with the automobile.

And a single bag of garbage reflects our jobs, our diets, our health, our hobbies, our habits. Just a quick peek reveals our eccentricities—even our neuroses.

At first glance, Sharon Zane's garbage is a reflection of the hectic schedule of the working mother. A Manhattan-based writer who uses a personal computer, she keeps a plastic trash bag in her West Side apartment exclusively for the paper debris of sentences that did not work. "By the end of the day I've created a mountain of paper," she says. And by the end of the day, the garbage bag in the study and the garbage bag in the kitchen are combined in an eloquent statement about the demands and dilemmas of the professional woman. On the days when the computer paper is particularly heavy, the bag that goes into the building's garbage compactor is filled with boxes of frozen fish cakes and Tater Tots.

The refuse of Ruth and Bob Magee of Massapequa underscores this retired couple's high-cholesterol diet of whole milk, butter, and eggs. The contents of the generic trash bags used by Joy and Harold Christophersen of Holbrook attest to the priorities of consumers who prefer bargains to brand names.

And the trash bags of Manhattanite Marianne Kavanagh are even more illuminating, if not all-encompassing. The bags include such dietary and socio-economic clues as Twinings tea bags, Pepperidge Farm cookie bags, and the pulp left over from fresh-squeezed orange juice. The sales receipts from Brooks Brothers and Saks are comments on her tastes. What is missing also matters—there are no cigarette butts or liquor bottles.

And how a person parts with what is no longer needed can reflect quirks of personality. Just imagine Felix Unger's garbage can.

Marianne Kavanagh dabs her mouth between sips of cappuccino as she confesses the deep, dark secrets that surround her garbage. "I gift wrap everything," she whispers. Marianne Kavanagh puts her tea bag in a Baggie and ties it up before she throws it away. She tears anything that has her name and address on it into tiny pieces. "Maybe I'm paranoid," she suggests. "Maybe I'm neurotic. No, don't say I'm neurotic." Every night after the 11 o'clock news, Marianne Kavanagh peers out the door of her 28th-floor apartment to make sure no one is in the hallway. She slips her bathrobe on over her nightgown and, if the coast is clear, she hurries to the compactor room with her small bag of garbage. She says she sleeps better knowing the garbage is out of the house.

Eccentricities can border on obsessions. Sonia Rodriguez of Brentwood washes out her soup cans and her husband, Juan, takes his concern for dental hygiene to an extreme. "He's a teeth fanatic," she says, in an effort to explain the preponderance of used dental floss, toothpaste, and mouthwash containers in their garbage—and the paucity of gum wrappers and junk food.

And one Suffolk woman, who asked that her name be withheld, tells tales on her garbage-obsessed husband—good-natured tales that might not be out of place in the "Believe It or Not" of Garbage World.

"In my house, if you're cracking peanuts, you'd better have a piece of paper handy, too, for the shells, because you can't just throw them in the garbage can," she begins. The rest of the rules read like the commandments of a man with a passion for order and cleanliness: Food garbage is not allowed in the bathroom or bedroom wastebaskets; broken glass must be wrapped in two paper bags; all paper boxes must be ripped at the seams. Even grease has its proper place.

"If you use a spatula to scrape grease or fat from a pan," the woman explains, "you'd better not wipe the spatula on the edge of the garbage bag. There's hell to pay for that one. Everyone knows fat belongs in the middle of the can—because if it's near the edge someone might walk by and get grease on his clothes."

The obsession with garbage as dirt goes hand in rubber glove with the commercial credo of "never touch garbage again." Jorie MacKinnon, who helped organize a recent recycling experiment by 100 East Hampton families, tells of one woman's nose-to-nose confrontation with raw refuse: "She'd always used plastic bags with ties, and here she was for the first time throwing food garbage in an unsealed paper bag. She called and said she had a problem. 'What's wrong?' I asked. She sounded kind of embarrassed. 'It's the garbage, it, it, it—well, it smells.' And I thought, we're so removed from our garbage—we even forget that it smells."

Both experts like William Rathje and socially conscious garbage producers like Jorie MacKinnon believe this attitude captures the essence of the mess—for the nation as well as the individual.

"Americans have an assembly line attitude toward garbage," Rathje says. "We use something, we throw it away, and it goes down the line to someone else. We think it's gone, but it's not gone. We simply push it away and don't think about it any more."

By contrast, people in other nations find ways to get the most out of their garbage. The Japanese give their newspapers and magazines to door-to-door entrepreneurs in exchange for rolls of toilet paper. In India, people make fertilizer by boiling animal bones they have collected from the garbage piles. The Dutch have created the world's largest compost pile—turning a million tons of refuse a year into approximately 125,000 tons of organic material for farm and garden usage. In Sri Lanka, man and beast alike wait at the dumps in Colombo for the garbage trucks to arrive— the human scavengers rush forward first, then the cows, pigs, and goats are allowed to feed on the heaps. And in Vietnam, one of the poorest countries in the world, there is virtually no litter because street beggars find a use for almost every scrap.

"Other societies see things as cyclical," Rathje says. "There is no end of the line; what goes around comes around. Garbage is always out there affecting something."

MacKinnon puts it on a more personal level. "We have to stop thinking of our garbage as kaka-poo-poo."

The Sigmanns' trash joins the assembly line the moment it goes into the plastic bag in the kitchen. As Deneen Sigmann observes, "I don't have time to fool around with garbage." They just want to get it out of the house. But the plastic bags sprinkled with canned corn and caramel apples stay tied to Garbage World—they tell the story of an American family.

The Sigmanns allowed a dozen bags of their garbage to be studied over the course of a month. The bags contained slightly more than 165 pounds of garbage. Deneen and Walter Sigmann and their four children throw everything—from spaghetti sauce to Sunday newspapers to insulin syringes—into the trash bag in their kitchen. By weight, the most prevalent items followed national patterns. The family discarded 42.6 pounds of newspapers, 42.2 pounds of food waste, 32.4 pounds of paper goods, and 19.7 pounds of plastics, among other items such as fabric and vacuum lint.

The bags offered material proof of Rathje's underlying premise: "We are what we throw away."

Their garbage showed that the Sigmanns go to McDonald's, they drink regular coffee and whole milk, eat pizza and takeout deli food. They read newspapers and own a cat and a dog. They try to save money by eating Hamburger Helper, bottom round, and chuck steak instead of better cuts of meat, but they gorge on frozen food and store-bought baked goods, which tend to be expensive. They spend small fortunes every week on packaged cakes and cookies, they eat Cheerios and cornflakes and the whole range of sugary cereals: Count Chocula, Trix, and Fruity Pebbles, to name a few. The family drinks almost one gallon of milk a day and eats about three loaves of bread a week. The younger children went trick-or-treating, and the family went shopping at a discount outlet mall in Pennsylvania. Walter Sigmann smokes a pipe, Deneen smokes menthol, filter-tipped cigarettes—about two packs a day. Deneen reads romance novels, and Jennifer wears Avon lipstick. Cathy is a Girl Scout and gets Bs in spelling and As in math. Mary writes home regularly from Drake University—begging her mother for care packages of her favorite store-bought chocolate chip cookies. Waltie changes the oil in the family cars and replaces the filter in the microwave oven. The Sigmanns are cable television subscribers, and they snack on pretzels, potato chips, and popcorn. They drink diet soda and throw away the twist-off caps but return the plastic bottles to the store. They attend Good Shepherd Roman Catholic Church, and, like many Americans, they do not even open the volumes of junk mail that arrive every day.

The Sigmann garbage can underlines Rathje's theme of truth-

One Family's Trash

Over the course of four weeks, Newsday collected and analyzed 12 bags of garbage from the six-member Sigmann family of Holbrook. The garbage added up to 166 pounds, the equivalent of an average-sized man. Here is what the Sigmanns threw out:

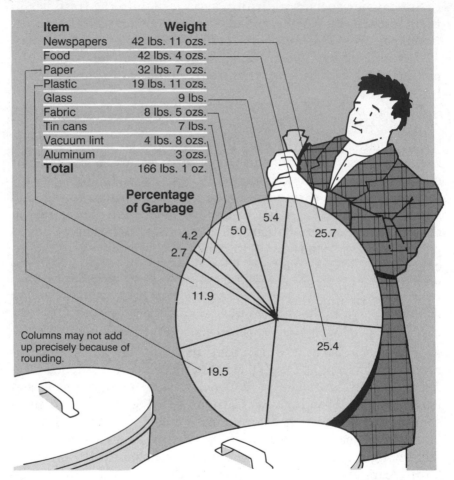

Item	Weight
Newspapers	42 lbs. 11 ozs.
Food	42 lbs. 4 ozs.
Paper	32 lbs. 7 ozs.
Plastic	19 lbs. 11 ozs.
Glass	9 lbs.
Fabric	8 lbs. 5 ozs.
Tin cans	7 lbs.
Vacuum lint	4 lbs. 8 ozs.
Aluminum	3 ozs.
Total	166 lbs. 1 oz.

Percentage of Garbage

5.4
5.0
4.2
2.7
25.7
11.9
25.4
19.5

Columns may not add up precisely because of rounding.

in-trash. It reveals their medical and financial histories—their prescription drugs, their credit cards, their bills.

And in many instances, the truth encompasses social exchange and ritual. In Manhattan—where the Islip garbage barge became a tourist attraction and people steal metal litter baskets to use as barbecue pits—trash can be part of the absurd circus of urban life. Patricia Kravitz takes her five-year-old daughter, Sara, to watch the garbage trucks unload the giant metal dumpsters in the alleyway of the West Side high-rise where they live. For New Yorkers who prefer the charm of brownstones to the convenience of elevator-buildings, taking out the trash means walking down—then back up—three, four, maybe five flights of stairs. Bags of garbage wait outside apartment doors until it is time to do such urban errands as moving the car to comply with alternate-side-of-the-street parking regulations.

In the suburbs, where homeowners wheel their molded plastic cans to the curb on metal trolleys, garbage becomes a way of socializing. On summer evenings, Carol Masak, who has lived in the same East Northport development for 29 years, chats with the man across the street as they both put out the trash. Sometimes they even share their garbage—when Masak has more trash than the carter will take, she carries a couple of her bags across the street.

The way America gets rid of its garbage has forced some communities to take the first steps toward recycling. From North Hempstead to Islip to East Hampton, town officials are grappling with new problems—like how to get homeowners to throw glass in green cans and paper in purple cans and food waste in cans of a different color. But most of the time, the American way leads only to de facto recycling. Carol Masak rolls newspapers into fire logs and puts coffee grounds and eggshells around the shrubs in her front yard. Sonia Rodriguez sends magazines to her sister in Puerto Rico. And one humid afternoon last summer, Patricia Kravitz performed what she considers an act of creative recycling.

For two weeks, she walked around Manhattan with an aluminum Coke can in her pocketbook. She felt guilty just throwing it away. And the five-cent deposit wasn't worth the ten-minute wait at the bottle-return section of her local grocery store. The way out finally appeared in the form of a homeless person picking soda cans out of a trash basket. Patricia Kravitz decided that tossing away her soda can served a greater good. "It was like seeing a remarkable chain," she recalls. "There are people in New York who are living off our soda cans."

But such informal stabs at recycling are easy because they do not require changes in lifestyle. They do not require us to break the habits that result in the flood of recyclable items now inundating our landfills. Each year, the average U.S. household discards 1,800 plastic items, 13,000 individual paper items, 500 aluminum cans, and 500 glass bottles.

Deneen Sigmann worries that the mechanics of recycling would produce separation anxieties. "I guess separation and recycling are probably good ideas," she says, "but I'm not going to bother until I absolutely have to. When somebody knocks on my door and says you have to do this or you go to jail, then I'll say OK, I'll do it. I won't like it but I'll do it. . . . It would be annoying. It would be an inconvenience. It would be time consuming."

The agonies and the ecstasies of recycling are related by Jorie MacKinnon in a journal she kept during the East Hampton experiment. She did not know what to do with the tiny staples that hold tea bags together, she questioned whether wine-bottle corks were biodegradable. Her husband, Steve, scrubbed melted cheese off the aluminum foil from a takeout order. And she worried about brunch guests. "Will we be able to keep guests from misplacing their leavings?" she wrote. She felt guilty at her job with an environmental group because she could not rinse the mayonnaise off her sandwich wrapper before throwing it away. She made a note to warn the cleaning woman about "the new system."

But it was not all kaka-poo-poo. She uncovered a secret soul of origami in a dismantled takeout Chinese-food carton, and she saw "terrific videos on recycling." She found herself "filled with the fervor of a recent convert." And she discovered a philosophy. "It's nice in a way to be developing a more involved relationship with our garbage."

But experts like Rathje believe most Americans just do not want to be bothered. "We haven't had to be as careful as some countries," he says, "but we're going to have to start conserving if we want to maintain the lifestyle we have today. We have the potential to change before it's too late."

It is a matter of bad habits being set in the concrete of everyday life. When it gets down to garbage, America seems caught in its own leftovers—mired in a mind-set exemplified by the Cinch-Sak slogan and Deneen Sigmann's exhortation to "just get it out of here." Garbage only becomes visible when it can no longer be ignored—when the bag overflows, when landfills dominate the horizon, when the Islip barge has no place to go.

And even then, it is difficult to look at the refuse teeming on our shores. Or to learn the lessons of our own history. "Garbage is part of life," Rathje observes. "Whether we want to see it or not, it will always be there."

In Deneen Sigmann's kitchen, the Cinch-Sak fills up without anyone in the family even noticing. In America's landfills, the markers of our time and culture accumulate. Oxygen cylinders, axes, and boots litter Mount Everest, and old lunar module parts fill the craters of the moon. Even the final frontier—space—has been polluted by the discards of a disposable society. More than 5,000 pieces of American garbage float in Earth's orbit—dead satellites, spent rocket stages, boosters, a power wrench, even a Hasselblad camera left behind after a manned mission.

Jorie MacKinnon offers a prophetic warning. "Without question, anything that you just don't want to know about is going to come back and get you."

CONTAINING THE CONTAINER

By Ford Fessenden

In the last 30 years, a dramatic change has taken place in the American garbage can.

In 1960, the largest component of U.S. garbage was organic waste—food, grass clippings, leaves. Now, the largest component is packaging.

Americans used to order meat from a butcher. Now, meats are individually wrapped in plastic film and Styrofoam or frozen in plastic "microwavable" containers. They once bought nails at the hardware store in bulk. Now, the nails are prepackaged in plastic containers.

That helps explain why Americans are the most prolific discarders in the world—and why it is becoming harder to find ways to get rid of their garbage. Food and yard waste break down naturally in dumps. But plastic packaging takes up space in a garbage dump, does not decompose for centuries, and can produce toxic gases when burned.

Unwilling to countenance the kind of effort to reduce or recycle trash that other industrialized nations have adopted, the United States has created so much packaging that it is beginning to choke on it.

In many instances, the container has overshadowed its contents. Packaging waste has increased 80 percent since 1960. A third of all garbage in this country is packaging, and more is being created every year. Americans throw out 400,000 tons less food than they did in 1960, a trend attributable to better processing and more convenient packaging. But that kind of convenience has come at a price—20 million more tons of packaging waste per year now than in 1960.

In fact, consumers paid more for food packaging in 1985—$29 billion—than farmers were paid to produce its contents, according to the U.S. Department of Agriculture.

And the trend is likely to accelerate in the future. Driven largely by the trend of two-earner families and the demand for convenient, single-serving processed foods, the market for food packaging is expected to more than double between 1985 and 1995.

Since World War II, the packaging of carbonated beverages has become a Goliath of throwaway containers. Nonrefillable bottles and cans were virtually nonexistent in 1947, and, as recently as 1965, Americans consumed three-quarters of their soft drinks from refillable containers that were reused, on average, more than a dozen times each.

Now, the situation is reversed. Less than one percent of the containers purchased by the soft drink industry in 1984 were refillable. Most of the 42 billion nonreturnable, nonrefillable containers ended up in the garbage: Even with recycling, carbonated-beverage trash has tripled since 1960.

The change is exemplified by packaging at a typical fast-food restaurant. Several years ago, the leftovers from a meal at such a restaurant were likely to include a paper wrapper from the hamburger, a paper soda cup, and a paper envelope from a package of fries. Now the meal probably includes a Styrofoam container for the hamburger, a paperboard box for chicken nuggets or fries, and plastic cups for sauce as well as soda.

The trend has done more than just create more garbage: It also has made it harder to dispose of. Plastics—the fastest-growing segment of the packaging boom—are difficult to recycle and take up at least three times as much landfill space for their weight as other materials, said Gretchen Brewer, recycling director for the Massachusetts Division of Solid Waste. The amount of paper in American garbage has also overtaken food and yard waste. But it is easier to recycle and dispose of paper than plastics.

Plastics could become a greater problem in the coming rush to incineration. Millions of pounds of polyvinyl chloride—a plastic compound that is known to corrode incinerator walls—are expected to be added to the garbage pile if the Food and Drug Administration allows its use as a food packaging material. Environmentalists and the New York City Department of Sanitation oppose the move.

Laws requiring deposits on bottles and cans have proven that it is possible to cut drastically the number of containers thrown on the trash heap. Although New York's 1983 law has not brought on a return of refillable bottles, it has at least encouraged recycling and has removed most of the bottles and aluminum cans from the landfills. But, although they have been proposed in most states and by Congress, only nine states have enacted bottle bills. Most legislation aimed at reducing packaging waste has been successfully contested by bottlers and packagers.

In Minnesota, a ban on the plastic milk containers that were replacing refillable glass in the '70s was immersed in a lengthy court battle by bottlers. State environmental officials won after six years in court and hundreds of thousands of dollars in costs, but by that time, the legislature had lost its taste for the confrontation and repealed the law.

In the early 1970s, when President Richard Nixon called for a "total mobilization" on environmental issues, a national bottle bill was introduced. But packagers mobilized, and the bill was never passed. Companies including Pepsico, General Foods, and the Aluminum Corporation of America helped found the National Center for Resource Recovery, which tried to promote high-technology processing of garbage and recycling as an alternative to limits on packaging.

"The only thing they wanted was no bottle bill," said Bill Kovacs, an environmental lawyer in Washington who was chief counsel to a House subcommittee dealing with the issue.

David Sussman, then a staff member of the federal Environmental Protection Agency, accused the group in a memorandum in the early 1970s of trying to stop cities from passing waste-reduction laws. He said the EPA, by awarding consulting contracts to the group, was "like the mayor taking a whore to church." The group has since expired.

In Europe and Canada, by contrast, initiatives by government have helped to stop the postwar trend toward nonrefillable containers.

Norway has a system of deposits and high taxes on throwaway containers. In the Land of the Midnight Sun, the throwaway bottle is virtually unknown—99.5 percent of beer and soft drinks are bottled in refillable containers.

PART II

LANDFILLS:
IS YESTERDAY'S SOLUTION TODAY'S PROBLEM?

CHAPTER 3

LANDFILLS HEAD FOR THE SCRAP HEAP

By Ford Fessenden

In 1984, the state of New Jersey abruptly closed a giant landfill serving Philadelphia.

During the next three years, the garbage dilemma of America's fifth-largest city would read like an existential drama: One by one, its choices would be peeled away, leaving it with no exit from an escalating crisis.

Suburban communities fought off every effort to locate a new dump to handle much of the 6,000 tons of garbage produced in Philadelphia every day. Ash from the city's incinerator was rejected by three states, the Bahamas, the Dominican Republic, Honduras, Panama, and the West African nation of Guinea-Bissau.

In 1987, for the first time in the city's history, the cost of garbage disposal exceeded the cost of fire protection, and Philadelphia was politically polarized over what to do next.

"Since 1985, we have tried everything," said Harry Perks, streets commissioner for the city. Philadelphia trucks its garbage to dumps as far away as Ohio, but residents there are up in arms.

"Nobody wants trash in their neighborhood. It does not smell nice. It does attract vermin, trash trucks, garbagemen. I see a potential that, even at any cost, you can't do anything."

The Philadelphia story is a harbinger for Long Island, New York City, and, eventually, the entire country. America is running out of space to bury 230 million tons of garbage every year.

Americans produce more garbage per person than ever—much of it elaborate packaging material and voluminous amounts of paper that did not exist 25 years ago. They have failed to respond to efforts to reduce the amount of waste produced or to recycle it. Worried about environmental threats posed by landfills, they have grown increasingly resistant to proposals to dig new garbage dumps.

"The sources of the crisis are in some ways very simple," said Michael Herz, staff attorney for the Environmental Defense Fund, in testimony before Congress in 1987. "There is more garbage . . . and fewer places to put it."

A *Newsday* survey of 56 states and territories found that more than 2,000 landfills have closed since 1982 for environmental reasons, and another 700 have closed because of lack of space. Hundreds of the remaining landfills do not have operating permits; at least 350 more are under orders to close.

By 1995, the U.S. Conference of Mayors fears, half of America's cities will have to close their dumps, and many will find it impossible to site new ones. Virtually every state in the union has a shortage of dump space, according to the Association of State and Territorial Solid Waste Management Officials.

The state of Washington had 260 landfills in 1984; it had just 100 by 1987. Minnesota has dropped from 160 to 100; New Jersey has gone from 350 to 59 since 1977. New York City is down essentially to one huge landfill—and a record increase in the amount of garbage dumped there has cut its life expectancy from 14 years to 10.

The crisis is worst in the Northeast—particularly in Philadelphia, New York, New Jersey, and Massachusetts—where the traditional "dump on the edge of town" is disappearing because sprawl has subsumed the edge of town. New York City and Long Island are among those in the worst predicament, their limited landfills full or filling rapidly. In 1985 and 1986, 83 landfills

The State of Waste Disposal
What states are doing about it, and which have the biggest problems

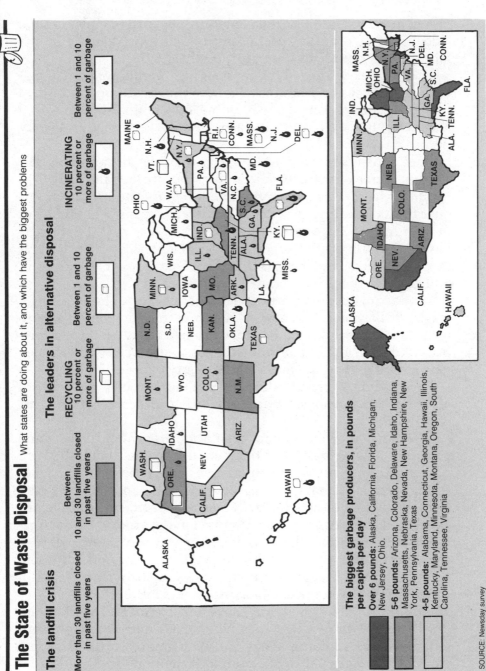

The landfill crisis

More than 30 landfills closed in past five years

Between 10 and 30 landfills closed in past five years

The leaders in alternative disposal

RECYCLING
10 percent or more of garbage

Between 1 and 10 percent of garbage

INCINERATING
10 percent or more of garbage

Between 1 and 10 percent of garbage

The biggest garbage producers, in pounds per capita per day

Over 6 pounds: Alaska, California, Florida, Michigan, New Jersey, Ohio.

5-6 pounds: Arizona, Colorado, Delaware, Idaho, Indiana, Massachusetts, Nebraska, Nevada, New Hampshire, New York, Pennsylvania, Texas

4-5 pounds: Alabama, Connecticut, Georgia, Hawaii, Illinois, Kentucky, Maryland, Minnesota, Montana, Oregon, South Carolina, Tennessee, Virginia

SOURCE: Newsday survey

Newsday/Steve Madden

closed in New York State, and only three opened to replace them. The state has ordered Long Island's landfills, situated atop the Island's only source of water, to close by 1990.

These problems have brought on a westward migration of garbage, a solid waste Manifest Destiny that is filling landfills in western Pennsylvania and spreading into Ohio, Michigan, and Kentucky. But even the far-flung landfills will reach capacity within a few years.

Once considered safer and smarter than the air-poisoning garbage incinerator it largely displaced in the late '60s and early '70s, the dump has become an environmental four-letter word. Poisons ooze from old dumps into drinking water. Dozens of landfills full of seemingly benign refuse, including the one at Old Bethpage, have turned up on the Superfund list of America's most dangerously polluted land because solvents, batteries, and paints in household garbage have contributed toxics.

More stringent environmental regulations have closed many of the landfills, and made new ones harder and more expensive to open. But the real crunch has been brought on by the "not-in-my-backyard" (NIMBY) syndrome, which has stymied efforts to open new landfills, even in places where plenty of space remains.

The syndrome refers to the ability of local groups, aligned with environmental groups, to stop the siting of landfills near their neighborhoods.

New York State's stringent regulations for landfill siting and pollution control make it difficult enough to site a landfill. Much of the state's land is ruled out because regulations prohibit landfilling over a primary source aquifer or in an area in which rainwater will run off the site. Practical considerations like hill slopes and poor roads eliminate more land. A study by Browning-Ferris Industries found that only about one percent of a 20,000-square-mile area that includes Long Island, New York City, and most of the eastern part of the state is suitable for a new landfill.

But that is when the decision really gets difficult.

"You go in and start doing some environmental work," said Ross Patten of Browning-Ferris. "People ask questions. You can't hide what you're doing. 'Why are we going to be the dumping ground?' They put political pressure on their public officials, who then pass a moratorium and zoning ordinances and resolutions, and 400 people show up at the county board meetings."

The Coming Landfill Crunch

How the number of active municipal landfills — meaning those still accepting garbage — is expected to decrease in New York State.

March, 1986

294

Capacity, Amounts Of Municipal
Waste Accepted Per Site
- 0 - 1,000 Tons/Yr.
- 1,001 - 10,000 Tons/Yr.
- 10,001 - 100,000 Tons /Yr.
- 100,001 - 500,000 Tons/Yr.
- Over 5,000,000 Tons/Yr.

January, 1996

76

January, 2006

13

The figures for 1996 and 2006 are based on projections of landfills closing because they are filled to capacity. The figures do not take into consideration the closing of landfills for other environmental reasons.

SOURCE: NY State Legislative Commission on Solid Waste Management

0   MILES   200

"It's a case of politicians who are elected for four years thinking in segments of that length," said Jim Staples of the New Jersey Department of Environmental Regulation, which is having trouble siting even recycling plants. "It's a matter mainly of political survival, to see if he can leave the problem for the next guy. The NIMBYs are well-organized, and it has a snowballing effect. People worry about pollution ... an ugly, indefinable something."

In Oregon, a series of public initiatives passed in referendums have frustrated Portland's attempts since 1982 to replace a landfill that will reach capacity in 1989, said Ernie Schmidt, senior solid waste engineer for the state Department of Environmental Quality. The city has begun talking of barging its waste 130 miles up the Columbia River to a new landfill.

In Florida, where the number of landfills has dwindled from 508 in 1974 to fewer than 200 in 1987, "it's become next to impossible to site a new landfill," said Ray Moreau, project manager for a waste-to-energy plant in Jacksonville. "In Florida, it's a combination of the public awareness of what has happened in terms of contamination from landfills, and the environmental sensitivity of this state. We have high water tables and a dependency on groundwater for 90 percent of our water."

Even in Texas, where there is plenty of space, the pace of landfill creation has slowed to a crawl. Hector Mendieta, director of the Division of Solid Waste Management of the Texas Department of Health, said new landfills used to be authorized at the rate of 25 a month, but local opposition has slowed that to about 45 a year. Austin has dropped plans to open a new landfill because of neighborhood opposition.

These capacity problems have seriously escalated the costs of dumping garbage. The average fee, according to a nationwide survey by *Waste Age* magazine, has begun an ominous climb—up 25 percent since 1985 to $13.43 a ton. And in some parts of the country, the rise has been stratospheric. Long Island towns are paying as much as $138 a ton to ship and dump their garbage hundreds of miles away, while, in Massachusetts, fees at some landfills have reached $140 a ton just to dump it.

The escalating costs for long-distance transport and dumping are warnings that the crunch is getting closer to the day that no one wants to think about—the day trucks do not pick up garbage because there is no place to put it down.

CHAPTER 4

THE HILLS OF FRESH KILLS

By Richard C. Firstman

The mountains of Fresh Kills are growing every day.

"See the World Trade Center in the distance?" said Ken Ustick, who was working in the marine-dispatching tower one day in 1987. The tops of the twin towers were barely visible over 130-foot peaks of garbage. "Next year, you won't be able to see it."

In 1948, Robert Moses, who was at the bottom of most things built at public expense in New York in those years, found a marshy 3,000-acre spot on Staten Island. He figured he would fill in the creeks and saltwater meadows with garbage for a couple of years, then stop, and then turn the landfilled dump into a park.

Things sort of got out of hand. Nearly 40 years later, the Fresh Kills landfill has evolved into the largest garbage dump on earth, worthy of an entry in the *Guinness Book of World Records*, a spot on the "Ripley's Believe It or Not" television show, and an occasional group tour conducted by the New York City Sanitation Department. And through four decades of garbage fashion—incinerators in the '50s and '60s, sanitary landfills in the '70s, and resource recovery plants in the '80s—Fresh Kills has remained a growing concern, even as it becomes an anachronism.

As the country moves toward the imported technology of

49

garbage-to-energy behemoths, Fresh Kills goes on and on—an immense final resting place for 94 percent of the residential detritus, retail discard, restaurant slop, and other undesirable elements produced by the people, places, and things of New York City. That comes to 44 million pounds every single day, on average.

That is a lot of garbage, but it tends to get lost in a 3,000-acre dumping ground, which is, of course, the idea. If the czars of solid waste ever devised a way to send trash into outer space and dump it on some unsuspecting planet, this is what the result might look like. A planet of garbage.

But within the next decade, even this vast garbagescape, built of garbage mountains and garbage roads, will reach what the garbage engineers define as a reasonable limit. By then, garbage and soil at Fresh Kills that are now 130 feet above sea level will have reached 505 feet—which would make it the highest point on the Eastern Seaboard south of Maine. It will be higher than the Washington Monument. And as the volume of garbage keeps growing, the life expectancy of the landfill keeps shrinking. Now, city officials say New York's main dumping ground will have to close within ten years. When that happens, the city hopes to fulfill its 40-year-old plan of turning the dump into a park—including a ski slope.

With trucks rolling in and out day and night, men in uniform walking around purposefully, and garbage-burying going on in four separate areas 24 hours a day, six days a week (sometimes seven), the New York City Sanitation Department's Fresh Kills landfill has the look of a military installation.

More than 600 people work at the dump, and an air of going-through-channels order pervades the place. The annual budget is $52.5 million. There are 281 pieces of heavy equipment—bulldozers, vacuum trucks, compactors, big garbage wagons, and the princes of heavy metal machinery, 90-foot cranes with gargantuan shovels to move piles of trash from here to there. It is Tonka Toy City for grownups.

The people who work there develop a certain affection for their landfill. They claim not even to mind the unprovoked attacks on the various senses. "After a while, you become immune to it," said Bill Aguirre, a 43-year-old sanitation worker, speaking of the smell of Fresh Kills. "My wife says, 'You smell like garbage!' but I

Cranes remove garbage from barges at Fresh Kills.

don't smell it." The Sanitation Department, which employs its own chemist, is always experimenting with deodorants.

"Actually, fresh garbage doesn't smell too bad," said Mike Massi, director of landfill engineering. "But if you excavate five-year-old garbage, there's a tremendous smell. We use that to test our deodorants." These days, pine oil is sprayed over the garbage after it arrives at Fresh Kills.

The lords of Fresh Kills are in the unenviable position of being bad neighbors, and knowing it. People who live, work, shop, or drive nearby talk about the landfill as if it were some awful family with nine cars on cinderblocks in the yard and screaming, naked children running around wildly.

"They have these disgusting barges sitting there for hours, especially on the weekends, and if it's hot, the garbage rots and it stinks," said Maxine Spierer, chairwoman of Community Board 3, which includes Fresh Kills. "They have done a lot to make it a less obnoxious neighbor. . . . But the basic problem is that you have a landfill."

And landfills produce even more objectionable secretions than odor. At Fresh Kills, a million gallons of leachate—the gunky residue of decomposed garbage picked up by rainwater—is released into the ground every day, polluting surrounding creeks. "The leachate is fairly high quality," said Dan Millstone, the Sanitation Department's environmental lawyer. "But it is something we're studying."

For years, the landfill's managers also have been dealing with the problem of escaped garbage—debris that seems to jump off boats and trucks and then heads for the hills, holing up in places like New Jersey, which is just across a narrow waterway.

Since 1979, the New Jersey township of Woodbridge had been arguing in federal court that garbage-at-large had been swimming to freedom and coming ashore at the town's beaches. Finally, in November, 1987, New York City agreed to spend millions of dollars for such things as a shield around the landfill and new cranes for precision unloading. Even before that, the city had boats sweeping up debris in the Kills, and cleanup teams patrolling the West Shore Expressway for escaped garbage.

There are two basic operations at Fresh Kills, officially termed Marine Unloading (14,000 tons a day that arrive by barge at two docks) and Truck Fill (8,000 tons, much of it commercial garbage brought by private carters who pay a tipping fee to dump it at two sites). The relationship between the two operations is something like that between the Army and Navy. As Tony Lobat, the man in charge of the Muldoon Avenue Truck Fill, sees it: "We're completely different. Their system is antiquated. They have exposed garbage. It's unacceptable. I would never work there."

The Marine Unloaders respond that they have more exposed garbage because they have more garbage. In fact, they have more of everything—huge, rumbling machines to transport the garbage from the dock to the dump, their own hangar-sized machine and electrical shops to fix them—and they seem to regard the Truck Fillers as minor leaguers. "It's like abortion and anti-abortion," Truck Filler Lobat said. "There are very strong feelings on both sides."

On the marine side, timing is everything. There is a constant flotilla of garbage on the high seas of New York, and the flow must be smooth and constant, lest the system become backed up. The trip begins in the collection trucks on the streets of the city. The trucks go to one of nine transfer stations on the banks of waterways in the boroughs, where the trash is dumped onto barges that hold more than 600 tons apiece. On a typical day, 20 of those barges are tugged to Fresh Kills, their journeys guided by

Newsday/Audrey C. Tiernan

As garbage trucks tip their load, bulldozers crush the mounds of trash to make room for more.

dispatchers in a tower overlooking the dock who must keep track of 100 barges and 5 tugboats at a time.

At the dock, crane operators scoop the garbage out of the barges and into pairs of wagons that are then pulled a mile or so on a garbage road to the "active face," or dumping area. The wagons are tilted, the trash is dumped. A bulldozer spreads the garbage, and a compactor rolls over it a few times. Later, soil is spread over each layer of trash.

Up on the Truck Fill, they use the direct-deposit method. Trucks drive up to the dump site, turn around and then, guided by landfill workers, back up and dump. Tony Lobat is known as a man who does not appreciate drivers who dillydally on the dump site. "They're here to dump and get off the landfill," he says. "We keep things on a law-and-order basis. You come in, you dump like a gentleman, and you leave."

Dumping like a gentleman can be an unseemly business. When a garbage truck arrives at Fresh Kills and raises its back end, an ugly mess comes oozing out: smashed green peppers, champagne bottles, shampoo, apple cores, vacuum cleaners, creamed corn boxes, stuffed animals, detergent boxes, Pampers.

Layer upon layer, the garbage is buried. After 40 years of this—1987 garbage on top of 1978 garbage on top of 1963 garbage—the mountains get pretty high. And it makes it hard to find things like murder victims, as Fresh Kills officials must explain to police detectives from time to time. "They get a tip that a body was dumped in a dumpster," Ken Ustick said. "But it's impossible to find anything. One time, a whole family came from Chinatown. The grandmother had thrown out $10,000. There were ten people looking for it. You try to isolate it. If you know where and when it was picked up, you can figure out what truck, when it was on the truck, which MTS [Marine Transfer Station], which boat, when it was unloaded. We can find you garbage down to within 1,000 pounds."

The happiest creatures at Fresh Kills are the gulls, who number in the tens of thousands at peak season, before they move south. "They eat cafeteria-style," Massi said. "And after they eat, they sort of loaf around."

Vic Carpentier, a sanitation worker, marveled at the charmed lives of the landfill gulls. "For them, it's one big constant smorgasbord."

GROUNDWATER THREAT UNCLEAR

By Thomas J. Maier

When state law forcing Long Island's landfills to close by 1990 was passed in 1983, it was embraced by the Island's entire delegation, signed by Gov. Mario Cuomo, and hailed by environmentalists as an effective weapon in the fight to save the region's fragile underground water supply from pollution.

At that time, few recognized the massive cost implications of the new law—forcing Long Island today to spend $150 million a year to truck garbage to distant landfills and accelerating the spending of more than $750 million for new incinerators. In the next few years, garbage disposal fees are expected to more than double in many Long Island communities.

Despite the heavy investment in incineration and long-hauling, however, there remains no evidence that pollution from municipal landfills has yet reached the deep underground wells that supply most of Long Island with its drinking water. Experts are still debating whether the 1990 deadline set by the law was justified by the extent of the immediate threat to the water supply.

"I think the reasons for the law were overstated," said Aldo Andreoli, director of environmental quality for the Suffolk County Health Services Department. "The public perception was that landfills would have an adverse impact on the entire drinking water system. And that's not correct. It was simplistic."

The deadline has been a major factor in a decision-making process that town officials describe as rushed. Long Island Regional Planning Board Executive Director Lee Koppelman said that, when the law was passed, the state forced towns into making quick decisions about building garbage-burning plants—without providing technical or financial advice. "The problem is that communities don't have much choice. The state says close the landfills, so it's ship it off the Island or find some form of incineration," he said.

Experts say landfills are only part of a much greater problem. The long list of suspected polluters of the Island's underground water supply ranges from leaky gas station tanks to pesticides and the use of oils and industrial chemicals. State studies show that several of the Island's approximately 27 active and recently

NOTE: Jane Fritsch contributed to this story.

closed public landfills have caused nearby groundwater pollution— forcing some shallowly dug wells to be closed. And, as water moves through the aquifers, studies have predicted that some landfill pollution will eventually flow into the water tapped by the main public wells.

But so far, landfill pollution has left unscathed those deeper wells, drilled hundreds of feet into the aquifers beneath the Island. "As far as we know, there is no landfill that is threatening a public water supply," said Al Machlin, the state Department of Environmental Conservation's regional engineer for environmental quality. "But the law does not allow us to let anyone pollute the groundwater— not just the deeper public water supplies."

Opponents now contend that the law was enacted based on a faulty premise and has accelerated the expensive move to incineration—creating a whole new threat

of air pollution. But even those opposed to the law for these reasons say it is politically difficult to support the notion of continuing to use landfills.

"There are four wells within a half-mile of the Hauppauge landfill, and they have yet to show any signs of pollution," said Islip Supervisor Frank Jones, whose town is building a $38 million incinerator. "But it's difficult not to close down the landfills. You tell people that there's no danger and they go crazy."

Supporters of the law say the already-polluted private wells near such landfills as those of Huntington and Smithtown underscore the eventual long-term threat to the Island's deeper aquifers, which supply drinking water. Without the law, they say, Long Island towns would not have made the crucial move toward waste-to-energy plants and garbage reduction.

"When the Legislature passed it

in 1983, they thought seven years was plenty of time for the towns to come up with the resource recovery plants," Machlin said. "None of the towns was going to spend millions of dollars unless they had to."

Sarah Meyland, director of watershed oversight and protection for the Suffolk County Water Authority, said the law was justified and necessary based on state and federal studies that found landfills a major source of pollution for shallow groundwater and an eventual threat to the deeper portion of the aquifers. "If we continue to landfill in the future the way we are doing now, it's only a matter of time before the pollutants pass directly into the main drinking water supply," she said.

Some critics contend that the law was passed without any cost-benefit analysis—weighing the environmental savings of closing Long Island's landfills against the cost of incineration and long-haul trucking. "There were letters written by environmentalists," recalled Gordon Boyd, director of the state Legislative Commission on Solid Waste Management. "And I think the [incinerator] industry felt that the law was good. You had NYPIRG (New York Public Interest Research Group) going around saying that water pollution from the landfills was going to kill everyone, and so everyone supported the law."

A co-author of the law, Assemblyman Lewis Yevoli (D—Oyster Bay), said the measure was crucial in trying to protect what was left of Long Island's unspoiled water supply. But Yevoli says he is appalled by the "astronomical" costs of long-hauling garbage from such towns as Oyster Bay. "No matter which way the taxpayer looks, the costs are horrendous with this problem," Yevoli said. "Those who say to wait until the water supplies are polluted don't know what they are talking about."

PART III

TRANSPORTING TRASH

CHAPTER 5

TRASH OVER THE LONG HAUL

BY WILLIAM BUNCH

Nearly ten million tons of U.S. garbage, traditionally buried in local town dumps, will be shipped hundreds of miles annually to rural landfills in a new $1 billion private industry that is largely unregulated by government.

Forced to close their own dumps, East Coast cities and towns are sending their garbage west at tremendous cost to local taxpayers and at the risk of spreading pollution. In some cases, the garbage is being hauled in refrigerated trucks or other trailers that are then used to ship food products back to New York, Long Island, and the rest of the Northeast—a practice health officials say could be dangerous.

The mass exodus of big-city garbage to landfills in Pennsylvania, Ohio, and Michigan is robbing rural communities of space to dump their own garbage in the future and raising fears about pollution. The high cost of trucking is increasing the pressure on garbage-exporting areas to build incinerators.

In 1986, Oyster Bay, Long Island, became the first town to ship garbage out of state. Town Attorney Robert Schmidt fears the consequences. "I have no doubt that . . . by 1997, we'll be shipping money to clean up the sites under Superfund legislation," he said. Hundreds of trucks from the Northeast are taking to the interstates every day. In 1988, the daily fleet swelled to roughly 1,350 tractor-trailers, each hauling 20 tons of trash.

"We're talking about a serious, major breakdown in a critical public service," said Gordon Boyd, executive director of New York's Legislative Commission on Solid Waste Management. Comparing garbage disposal to other public services like drinking water, roads, or fire protection, he said that "if any other system were performing as poorly as solid waste management, people would be in the streets. There would be revolution."

Homeowners in some Northeastern communities are paying as much as $240 a year in extra taxes to truck their trash to the Midwest. But officials in Pennsylvania, Ohio, Kentucky, and elsewhere are now adopting laws or guidelines aimed at stopping the flow of out-of-state garbage, leading to fears that the Northeast's garbage may have no place to go.

When Oyster Bay's landfill closed in 1986, it shipped its garbage 160 miles to the Scranton, Pennsylvania, area. A year later, it was forced to go as far as the woodlands of southern Michigan, 635 miles away, and the bluegrass hills of western Kentucky, an 850-mile odyssey, to find landfills that would take Long Island trash.

To cut costs at those distances, East Coast haulers are shipping out garbage on interstate trucks that come from the Midwest carrying consumer products, including foodstuffs, and are looking for a load for the return trip. A trucker from Rockland County, New York, hauled Long Island garbage to Ohio, steam-cleaned the trailer, and said he was headed to a Fort Wayne, Indiana, brewery to pick up beer.

The trucker, Brian Murphy, an owner-operator, said he sprinkles coffee grounds on the floor of the trailer after the cleaning to remove the smell of the garbage. "It's like an old wives' tale," he said. "You sprinkle coffee, it polishes up the job." Such tales are common in an industry so new that federal, state, and local environmental agencies have no specific laws or regulations to deal with it.

A six-month *Newsday* examination found:

- There are no laws to prevent the hauling of garbage in refrigerated trucks or other rigs that haul food products. Several health experts warned that, even with steam-cleaning of the trailers, this practice might spread disease and should be stopped, at least until sanitary tests can be carried out.
- Although the purpose of hauling garbage hundreds of miles is to prevent pollution, several Midwestern landfills that have accepted East Coast garbage have been cited for polluting water supplies or failing to keep up with burying the daily trash flow. For example, Oyster Bay's landfill is closed because of serious pollution problems that have placed it on the federal Superfund cleanup list. But of the landfills that have accepted Oyster Bay's garbage, two in Pennsylvania and one in Ohio have been closed, while a Michigan dump is under investigation as a possible Superfund site because of groundwater pollution.
- To dump garbage in the Midwest, East Coast homeowners are paying enormous tax bills. In Oyster Bay, town officials say an average homeowner paid $450 for garbage pickup and disposal in 1988—roughly double what residents of neighboring towns on Long Island paid on average to get rid of trash.
- Towns that truck their garbage to distant, privately owned landfills are risking lawsuits if those landfills later must be cleaned up, many lawyers believe. Islip Town, in rejecting a plan to contract for long-distance garbage hauling in 1986, cited a legal opinion that Islip could pay millions in damages if out-of-state landfills leaked toxic chemicals.

The push to ship an estimated ten million tons of trash a year—by truck, rail, and sometimes by barge—has moved ahead with very little advance planning. Through the 1970s, before the crisis in landfill space, American towns dumped their trash within their boundaries or close by. But by the turn of the decade, groundwater pollution from toxic chemicals became a major issue, and studies blamed municipal trash as a source of the pollution.

The problem proved to be particularly acute in the coastal areas of New Jersey and on Long Island, which depend on groundwater and have densely populated suburban areas. On Long Island, state law requires landfills to be closed by 1990, and in the meantime, state officials have forced towns such as Hempstead and Oyster Bay to truck garbage rather than expand dumps.

For many towns, the trucking is a short-term fix intended to handle garbage until incinerators can be built and operated. But trucking may be a long-term necessity because the incinerators produce large quantities of ash, and because they sometimes close for repairs and cannot process all the garbage their communities will produce.

On Long Island, little garbage was trucked away until 1984, when Hempstead closed one of its two landfills and began shipping about 50 truckloads a day to Orange County, New York, a journey of 95 miles.

By 1989, three towns with roughly half of Long Island's population—Hempstead, Oyster Bay, and Huntington—are expected to be trucking all their garbage away. This trash odyssey of more than 1.3 million tons a year will amount to roughly 220 truckloads a day. That means that, on average, one tractor-trailer loaded with 40,000 pounds of garbage would enter the Long Island Expressway every six and a half minutes.

In fact, garbage will become the region's chief export. Long Island will ship 12 times as much trash as its 1986 potato crop and a whopping 150 times more than its duck production. The town governments on Long Island have fought long-distance garbage hauling because of the high cost, but state officials have ruled that they have little choice because of a 1983 law restricting landfill expansions. And Long Island is not unique. New Jersey, where several giant landfills were closed or restricted in 1987, shipped out more than five million tons of garbage in 1988—more than half of all the garbage it produced. Philadelphia, which exports 1.5 million tons, and the Boston area, which trucks more than 800,000 tons, are sending garbage out of state because they have been frustrated by neighborhood opposition in their efforts to build new garbage-to-energy plants.

The high price of long-distance garbage hauling—as much as $138 a ton—has led to the widespread use of commercial tractor-trailers, which normally carry consumer products. The contractors who arrange to truck trash off Long Island and elsewhere have found they can ship more cheaply with "backhaulers"—tractor-trailers that haul Midwestern products into the populated

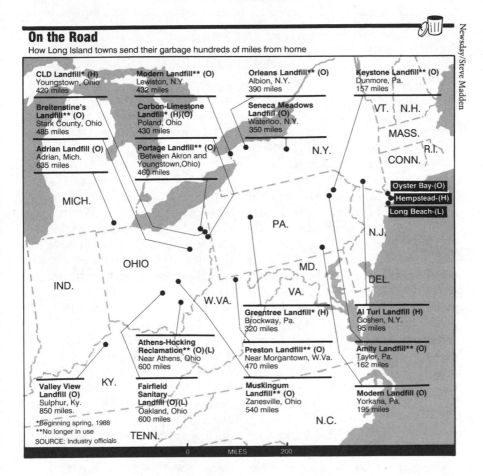

On the Road

How Long Island towns send their garbage hundreds of miles from home

Newsday/Steve Madden

CLD Landfill* (H)
Youngstown, Ohio
420 miles

Modern Landfill** (O)
Lewiston, N.Y.
432 miles

Orleans Landfill** (O)
Albion, N.Y.
390 miles

Keystone Landfill** (O)
Dunmore, Pa.
157 miles

Breitenstine's Landfill** (O)
Stark County, Ohio
485 miles

Carbon-Limestone Landfill* (H)(O)
Poland, Ohio
430 miles

Seneca Meadows Landfill (O)
Waterloo, N.Y.
350 miles

Adrian Landfill (O)
Adrian, Mich.
635 miles

Portage Landfill** (O)
(Between Akron and Youngstown,Ohio)
460 miles

Oyster Bay-(O)
Hempstead-(H)
Long Beach-(L)

Greentree Landfill* (H)
Brockway, Pa.
320 miles

Al Turi Landfill (H)
Goshen, N.Y.
95 miles

Athens-Hocking Reclamation** (O)(L)
Near Athens, Ohio
600 miles

Preston Landfill** (O)
Near Morgantown, W.Va.
470 miles

Amity Landfill** (O)
Taylor, Pa.
162 miles

Valley View Landfill (O)
Sulphur, Ky.
850 miles.

Fairfield Sanitary Landfill (O)(L)
Oakland, Ohio
600 miles

Muskingum Landfill** (O)
Zanesville, Ohio
540 miles

Modern Landfill (O)
Yorkana, Pa.
195 miles

*Beginning spring, 1988
**No longer in use
SOURCE: Industry officials

0 MILES 200

New York area and are looking for cargo to make money on their return trip.

To seek "backhaulers" to take trash from Oyster Bay and discards from private recyclers on Long Island, Pennsylvania-based truck brokers have advertised on computer networks for both regular tractor-trailers and refrigerated trailers—which are designed to haul food products.

Although several public-health and trucking-industry experts are concerned about the use of refrigerated trucks to haul garbage, officials with federal agencies such as the Environmental Protection Agency and the Food and Drug Administration, as

ON THE ROAD

By William Bunch

In his first month as an interstate trucker, Keith Ackerman hauled everything from women's clothes to Mexican bean dip, cruised the open roads from Plaquemine, Louisiana, to St. Paul, Minnesota, and got an enticing offer over his CB radio at a Pennsylvania truck stop from a woman with the handle "Lady Feelgood."

But nothing prepared him for the marching orders he received when he pulled his extra-wide, 48-foot-long trailer onto Long Island on a Monday afternoon and called his dispatcher in Indiana. "Compressed garbage?" Ackerman asked the dispatcher. "What the hell is compressed garbage? There's no money in compressed garbage."

"Don't kid yourself," the dispatcher answered.

Over the next 72 hours, Ackerman would get a firsthand introduction to the sights—and smells—of America's new $1 billion over-the-road industry: long-distance garbage trucking. And his journey with 23 tons of garbage crammed four feet high into every nook and cranny of his trailer would be an American travelogue of jukeboxes and truck stops—an interstate odyssey beyond graffiti-scarred subway stations in the Bronx, over the grassy peaks of western Pennsylvania, and past the American flags and Victorian front porches of Midwestern main streets.

It would be a costly trip for the taxpayers of the town of Oyster Bay—$115 a ton for a total of $2,645 for the load. When they plunked their plastic bags into curbside cans, they could have little idea that their refuse would wind up in an American heartland whose residents would greet the worst the Island has to offer with car blockades.

Ackerman expressed his own aversion to his cargo after his trailer was loaded at the Oyster Bay garbage transfer station at Old Bethpage. "I don't like hauling garbage," he said. "It's depressing pulling in there—the smell of it."

Normally, there is little that depresses Ackerman, a wiry, fast-talking 30-year-old who lives on his family's farm just outside of Toledo, Ohio. With two pictures of his best girl, Candy Sue, stuck up inside his cab, along with two packs of Winstons, a jar full of jawbreakers, cassette tapes ranging from Iron Maiden to George Jones, and his upside-down CB radio, Ackerman was ready for the uncluttered interstates where he can "hammer down" his accelerator.

Ackerman, who works as a

driver for a small Ohio firm that leases its nine trucks to the larger Whiteford Truck Lines of South Bend, Indiana, had spent the night in a sleeper berth in his cab at a Commack truck stop after hauling paper rolls from Marion, North Carolina, to Maspeth, Queens. He was at Old Bethpage at 8 a.m. Tuesday but had to wait until noon before he could back his trailer into the loading area, where garbage—baled like hay into one-ton blocks—filled the 408-square-foot floor of the trailer.

When Ackerman pulled his giant rig onto the Long Island Expressway Tuesday afternoon, it was doubtful that passing drivers had any idea his cargo was a mass of rotting leaves, old newspapers, shopping bags, and beer cans. Nor did those motorists—including several who cut perilously close in front of the rig in the late afternoon traffic—suspect what a heavy load he was hauling. The tractor-trailer and cargo weighed 79,600 pounds—just 400 pounds under the federal weight limit.

"I'm in no hurry," he said—a slogan that became a theme for the next two and a half days.

Ackerman broke up the monotony of the road with leisurely meals like the ham steak and French fries he ordered at a Union

76 truck stop in Pennsylvania, where he dropped his quarters in the jukebox, in the video-game machines, and in the pay phone to call Candy Sue.

On the highway, Ackerman, whose handle is "the Axeman," kept his CB radio tuned to the constant chatter of other long-haul truckers. The main topic was avoiding police speed traps and impromptu state inspection stations. "Eastbound at the 74-yard-line [mile marker]—you got a bear that wants to take your picture, so look real purty," one trucker warned his comrades on Interstate 80 in a slow, Southern drawl, referring to police.

For Ackerman, the main worry was not speed traps but scales. He paid $9 to weigh the rig for himself at an Emlenton, Pennsylvania, stop and found that the rear axle weighed 38,800 pounds—4,800 pounds over the limit. Even after he repositioned the wheels of the trailer, the rear axle was more than 1,300 pounds over its limit. That placed his company at risk of a fine by state authorities, who say that overweight trucks damage roadways and are difficult to stop in heavy traffic.

Just outside of Youngstown, Ohio, Ackerman pulled into the state's weigh station. But five large tractor-trailers were ahead of him, and the garbage-hauler got a green signal to pass on through.

"All right—I don't have to get weighed!" Ackerman said, flexing his right arm in a symbol of victory.

He spent Wednesday night at an Interstate 70 truck stop in Ohio called the Buzz Inn, which advertises, in large black letters, "BEER" and "AMMO." Then the rig lumbered through the narrow streets of Lancaster, Ohio, and around rural roads.

In Oakland, Ohio, a community of about 12 homes, a gas station, and a general store, Ackerman turned left at the only intersection, pulled his 40-ton rig across a bridge with a 26-ton weight limit, and passed handmade signs that read: "Don't kill our children with out-of-state trash."

His final stop was the 6-year-old Fairfield Sanitary Landfill, owned by Mid-American Waste Systems, in Oakland. In July, 1987, large numbers of trucks from Long Island and Philadelphia began arriv-

well as several state and local agencies, said they know of no law that specifically prevents the practice. A spokesman for the U.S. Agriculture Department said his agency would be concerned if meat or poultry were hauled in trailers used to ship garbage, but agency inspectors would have to be aware of a specific episode.

The majority of garbage trucked to the Midwest is trucked in regular tractor-trailers, but environmental groups in Ohio have noted a number of cases of trash-hauling in refrigerated trailers. In September, 1987, a group that monitored and videotaped the Fairfield Sanitary Landfill in Oakland, Ohio, for six days recorded 28 refrigerated trailers hauling garbage. An engineering audit of the Oyster Bay hauling operation in the summer, 1987, said a truck broker in Norristown, Pennsylvania, was arranging "for haulers who have brought a load of agricultural or manufactured products to the East Coast to pick up a load of [solid waste] at Oyster Bay and drop it off at [two Ohio landfills] on their way west." The audit said the trailers are steam-cleaned after dumping the garbage. Oyster Bay's garbage-hauling contractor said that a few refrigerated trucks made pickups that summer but that such rigs are now barred.

The Norristown broker, Jack Castanova, said some of the trucks he hires to haul Northeastern garbage are refrigerated trucks that also carry produce and meat, but he said they are completely disinfected with steam-cleaning before taking on new

ing—drawing protests from nearby residents, who fear a decrease in property values and the threat of groundwater pollution. More than 50 residents circled the landfill in cars and pickup trucks in an effort to blockade the out-of-state trucks from dumping.

"It's almost a religion to us," said George Bartram, who lives a few hundred yards down a rolling hillside from the dump and heads a newly formed citizens' environmental group. "We want our grandchildren to have a safe place to live. This was a nice neighborhood before this happened."

To bring Long Island garbage to Bartram's neighborhood, Ackerman had driven more than 700 total miles and burned roughly 118 gallons of fuel. The bumpy road into the dump site damaged his drive shaft. After trying to fix it Thursday afternoon, he abandoned the rig at the landfill gate and got a ride to Columbus.

"I'll send them the paperwork in the mail," Ackerman said. "I'm through with garbage."

cargo. He said refrigerated trucks that haul meat often smell bad from dripping blood, while "this stuff [garbage] is so safe that people have it in their kitchen."

"If we have an empty trailer, we try to fill the thing up," said Donald Nugen of Day's Express, an Indiana trucking firm that has hauled baled garbage off Long Island in refrigerated trailers. He said most of Day's Express's 200 rigs are refrigerated trailers that haul foodstuffs, such as produce, and general goods, but that trailers were steam-cleaned after hauling garbage. "There's no health problem at all," he said.

But Dr. George Kupchik, a public health expert and former chief engineer for the New York City Department of Sanitation, said the garbage cargo could include infectious matter such as diapers. "You really have to steam it out thoroughly, and I don't know if that would be done. This is a very crude business—it's not like the inside of an operating room. You have to get people to do it thoroughly, and that's painstaking and expensive."

Alan Franks, a spokesman for the Ohio Environmental Protection Agency, said garbage hauling in refrigerated trucks is a new phenomenon that concerns state officials. Said Franks: "At best, it's not a good idea. At worst, we're not sure what's hauled in—it could be hospital wastes—and the stuff going back to you guys in New York or Pennsylvania could be food."

The lack of regulation of long-distance garbage trucking is in sharp contrast to federal rules on hazardous-waste shipping. For toxic wastes, the EPA monitors shipments through documents called manifests that track the wastes from the factory to the landfill or disposal site. But an EPA official said a similar system for garbage is neither practical, because of the large volumes involved, nor necessary, because garbage is not as toxic as industrial wastes.

"We have generally looked at solid waste management as something the states and municipalities ought to be dealing with," said Joseph Carra, director of the EPA's waste management division. "They are the government entity ultimately responsible."

John Moore, spokesman for the New York state Department of Environmental Conservation, said, "Long-distance hauling of municipal waste is a pretty recent development. Until that started, hauling was mostly inside the community—within a town, within a county. So the state has not regulated it."

While state officials view long-distance trucking as a solution to Long Island's garbage crisis, the daily caravan of tractor-trailers has contributed to the region's traffic woes.

In August, 1987, a Day's Express tractor-trailer carrying garbage from the Jet Sanitation transfer station in Central Islip to Oakland, Ohio, split apart and spilled its cargo onto the Staten Island Expressway, snarling traffic for hours. A similar incident took place in November, 1987, on the Cross Bronx Expressway, when a tractor-trailer hauling garbage from Westbury Paper Stock to an Ohio landfill jackknifed and spilled nearly 45,000 pounds of refuse on the roadway, tying up traffic for five hours at the height of the evening rush.

Despite the environmental and roadway risks, perhaps the most immediate drawback of long-distance garbage trucking is the cost. Oyster Bay blamed garbage hauling for a 47 percent tax increase for 1988.

As other communities look ahead to trucking garbage, their taxpayers face sharp increases. In Hempstead, where the Oceanside landfill was slated to close during the summer of 1988, the town says the garbage bill for the average homeowner was expected to rise by $115 to $294. A similar increase—ranging from $60 to $120 a year to yield total bills of $210 to $270—is projected for taxpayers in Bergen County, New Jersey, where the county is now scrambling to find a contractor to haul its 3,750 tons of daily trash.

Some environmental and town officials complain that the nearly $1 billion a year price tag of long-distance garbage trucking for the nation is a waste of money that could instead clean up dozens of existing pollution sites, or build ten large garbage-to-energy plants each year.

"That money would clean up . . . most of the landfills on Long Island," said Guy Germano, the town attorney of Islip, which fought the state and, like Oyster Bay, won a landfill expansion to avoid trucking its garbage off Long Island. "That money, leveraged, could buy property to protect the groundwater or build schools and medical facilities—name the things we need."

Critics say the fact that trucking does not prevent pollution—it only moves it from one place to another—shows that the practice makes no sense on a national level.

In 1986, most of Oyster Bay's garbage went to the Amity Landfill in Taylor, Pennsylvania, but Pennsylvania state officials closed the landfill in December, 1986, saying the landfill was taking in much more trash than its daily capacity and that heavy trucks were damaging local roads. So, Oyster Bay's garbage was diverted to the Keystone Landfill in Dunmore, Pennsylvania. There, Pennsylvania authorities charged, the dump was accepting four to six times its permitted amount in daily trash from Long Island, New Jersey, and Philadelphia, not properly covering the garbage, and allowing polluted water to collect on the surface. State officials closed the Keystone Landfill in April, 1987, and fined its owner a record $500,000.

In May, 1987, Oyster Bay's trucking contractor, Willets Point Contracting Corp., citing the unforeseen shutdown of the two Pennsylvania landfills, invoked the so-called "force majeure" clause of its contract with Oyster Bay. That raised the hauling price from $70 a ton to as much as $138 a ton. The garbage would now go to sites near Buffalo and Rochester, as well as Ohio, Michigan, and Kentucky.

But one of the Ohio landfills, in Portage County, was closed by state officials who charged that polluted water—known as "leachate"—from the overburdened landfill was running into a nearby stream. The state officials described it in court as "a perpetual toxic leachate-generating machine."

One of the most distant landfills to accept Long Island's trash, in Adrian, Michigan, is in the process of being shut down by Michigan officials because of alleged problems with groundwater pollution and its location on a flood plain. "There's no bottom to the landfill," said Kay Brower, a Michigan state environmental official. "There's concern the landfill juices will wind up leaking out the bottom. It's an old site that was operating before the stringent [federal antipollution] laws passed in 1978."

Islip's Germano said the potential pollution problems at out-of-state landfills—and the multimillion-dollar cleanups that might result—were prime reasons that Islip did not want to truck away its trash. "We looked at some landfills that are not great landfills," he said. "Many of the landfills were unlined, and many were in

A tractor-trailer carrying Oyster Bay garbage approaches a landfill in Oakland, Ohio.

areas of sensitive water. We were looking at possible problem landfills in the future, and we did not want to become involved."

The public outcry against out-of-state garbage in Ohio and elsewhere is surging to the forefront in state legislatures, where lawmakers are studying ways to limit severely the flow of East Coast garbage. The lawmakers cannot ban out-of-state garbage outright. New Jersey tried to do that in 1978, when it was receiving most of Philadelphia's trash, but the U.S. Supreme Court ruled that states cannot restrict interstate commerce. But the new round of efforts by states to restrict outside garbage received a big boost in the fall of 1987, from a ruling in the Ninth Circuit of the U.S. Court of Appeals in a Portland, Oregon, case. It ruled that the Portland metropolitan government—to extend the life of its landfill—could restrict private haulers from nearby Washington State. Experts said that state legislative actions based on this court ruling could leave East Coast communities with nowhere to get rid of trash for the next few years while their garbage-to-energy plants are under construction.

In Ohio, Governor Richard F. Celeste signed a law in the summer of 1987 that authorizes counties, once they establish garbage districts, to decide if they want to bury outside waste. A similar bill failed to become law in Pennsylvania in 1988.

"It [the Oregon court case] sent a signal to Ohio, Pennsylvania, and other receiver states that if you draft the law properly, you can restrict New York waste from coming in—and relatively soon," said Boyd of New York's legislative commission.

In the meantime, the long-distance trucking of garbage has proved profitable for a network of landfill operators, truckers, and private contractors. Some of the major Midwestern dumps that accept out-of-state trash are run by local entrepreneurs, but a large amount of the refuse is trucked to landfills owned by major national garbage firms. Of the five landfills now accepting the bulk of Oyster Bay's trash, four are owned by large national companies—Browning-Ferris Industries, Waste Management, and Laidlaw Waste Systems.

Because they compete for business, these private landfill operators are reluctant to reveal their dumping fees. But in the face of the garbage crisis, several Ohio landfills have two pricing systems: one for local trash and a higher rate for out-of-state communities desperate for landfill space. Chris White, the president of the Fairfield Sanitary Landfill in Oakland, Ohio, refused to reveal the dumping fee for Long Island and Philadelphia garbage. But he said that, after accepting the out-of-state trash the summer of 1987, he was able to slash dumping costs for the nearby town of Lancaster, Ohio, by $100,000 for that year.

The national firms also are entering the hauling business. Browning-Ferris won a 13-year contract to haul Hempstead's garbage to Pennsylvania and Ohio until the town's garbage incinerator opens, and to truck away unburned garbage and the ash residue from the plant. The $852 million Hempstead contract is expected to bring BFI millions in profits.

While the $40 million annual cost of garbage hauling is the largest item in Oyster Bay's budget, town officials said they would not release information on how much the hauler, Willets Point Contracting Corp., is earning. They said that releasing the information would hurt the contractor's efforts to win new hauling contracts.

While Oyster Bay residents pay $115 a ton to truck their garbage to Ohio, the actual trucking cost is much less. A Pennsylvania truck broker quoted a shipping rate of $21 a ton to haul the trash. The rest of the money pays for the landfill space—usually in the range of $15 to $30 a ton—the truck broker, a subcontractor that runs Oyster Bay's garbage transfer station, and profit for Willets Point Contracting Corp.

Now entrepreneurs are promoting schemes to take advantage of the Northeast's solid-waste crisis by turning garbage into gold—with abandoned coal mines, railroad cars, and even with barges.

Pushing one of these ventures is a name familiar to many New Yorkers: Bob Tucker, a tight end for the New York Giants in the 1970s. Tucker now heads a subsidiary of a Pennsylvania coal company that wants to convert a 3,400-acre mining site in Luzerne County, Pennsylvania, into a gigantic waste facility that would include landfills, garbage-to-energy plants, sludge processing, and composting. To Tucker, the garbage facility is the best way to reclaim the land, boost the economy of the depressed coal region, and pump $150,000 a year in new tax money into town coffers. Said Tucker: "This could be the catalyst the area needs."

Like other landfill entrepreneurs, Tucker foresees the garbage arriving not only by road but by rail. Passaic County, New Jersey, has an ambitious plan to ship 1,300 tons of trash by rail each day to a landfill near Pittsburgh. However, the experiment so far has been derailed by Keystone State limits on dumping in that landfill, which has forced the contractor instead to truck the garbage to Ohio and West Virginia, as well as Pennsylvania.

And despite the Islip garbage barge fiasco, some businessmen still feel the only economical way to transport trash will be over the high seas.

In Bergen County, New Jersey, officials opened bids to haul away 3,750 tons of garbage a day. The only formal bid was from a Manhattan firm that would get rid of the trash for $92 a ton—by shipping it on barges to landfills in Panama and Costa Rica.

In Steubenville, Ohio, a New Jersey firm purchased 1,000 acres of property scarred by strip mining and plans to reclaim the land by burying garbage to be shipped in by rail—as much as 1,750 tons a day for the next 50 or 60 years. To make the proposal acceptable to local residents, the company has offered to accept Ohio trash at the bargain rate of $4 a ton while pumping an estimated $675,000 a year into a local foundation. The residents' response has been to carry protest placards like "Love New Jersey, Not Their Garbage" and "I'd Rather Fight Than Stink."

"They've come with absolutely nothing," said J. J. Bruzzese, Jr., the Steubenville attorney representing the landfill owner. "There's no objective reason [to oppose the landfill] that appeals to your head. There are a lot of things that would appeal to your gut."

In Athens County, Ohio, protesters have been successful at stopping out-of-state trash. Karen Harvey, an Athens County commissioner, said residents were outraged because the outside trash was filling up the local landfill while the county was in the second year of Ohio's most aggressive recycling program.

"People were seeing trucks with New York garbage, while we were washing out aluminum cans and stacking our newspapers," Harvey said. "Although the Ohio EPA tries to make landfills as safe as possible, we don't know what's going to percolate out of those things 10, 25, 50 years down the road. We don't know if Oyster Bay, Long Island, is going to take care of it."

WHERE WILL ALL THE GARBAGE GO?

By William Bunch

Since 1986, a cargo ship called the *Khian Sea* has quietly drifted through the South Seas in a voyage that has made the Islip garbage barge's odyssey seem like a Sunday afternoon rowboat ride in Central Park.

The *Khian Sea* beats the *Mobro 4000* in time, cargo, and distance. Islip took its prodigal barge home after a mere six months, but the ship filled with ash from Philadelphia's incinerators wandered the seas for 26 months before it mysteriously unloaded its cargo. The *Mobro* carried a mere 3,186 tons of garbage, but the *Khian Sea* was loaded with nearly 15,000 tons of ash. And while the garbage barge was banned all the way to Belize, the ash boat has been barred from places as far away as Guinea-Bissau in West Africa.

In November, 1988, port officials in Singapore announced that the *Khian Sea*, renamed *Pelicano*, had dumped the ash. But no one disclosed where.

"I do not know what they did with the ash," said a man who identified himself to reporters as Captain Arturo Fuentes.

The *Khian Sea* saga is just one episode in Philadelphia's garbage story. Without a landfill to its name, America's fifth-largest city has been depending for a decade on the kindness of strangers—New Jersey, Ohio, and other states and nations—to bury its incinerator ash. Those strangers have not always been kind.

"Reason and logic really don't have anything to do with this," said Harry Perks, the streets commissioner for Philadelphia, who has been frustrated at every turn in his efforts to find an affordable way to get rid of the city's ash.

Philadelphia's garbage crisis underlines the problems that many localities are facing in trying to dump their trash. These include rural opposition to urban garbage and concern that U.S. cities—shunned in their own back yards—will try to get rid of trash in Third World nations in what one Philadelphia editorial writer called "waste imperialism."

Most of the 6,000 tons of garbage produced each day in Philadelphia is trucked out of town to landfills—some as far away as Ohio. But the greatest uproar has been caused by the ash residue from the 1,200 tons of trash burned daily in two city incinerators.

In 1986, ash from the City of Brotherly Love was spurned by Maryland and Virginia. South Carolina finally accepted three bargeloads, but with a reluctance marked by its governor, who sent a sack of garbage to Philadelphia Mayor W. Wilson Goode.

Then a contractor for the city came up with the idea of sending ash to a landfill project in the Bahamas, and the *Khian Sea* set sail. Only the Bahamas did not want the ash, and neither did the Dominican Republic, Honduras, or the West African nation of Guinea-Bissau.

The latest setback for Philadelphia involved a plan to ship ash to Panama as fill for a new road to a planned resort complex. But just days before the first ship was to head out in September, 1987, the environmental group Greenpeace unfurled a giant banner from Philadelphia City Hall saying, "Don't export toxic ash to Panama." Panamanian officials—citing fears that the ash was toxic and would poison lobsters and other marine life—killed the proposal.

"There were two angles," said Jim Vallette, a Greenpeace activist who led the campaign. "One was the whole drive across the country to build incinerators—this seemed to be an important symbol of how incineration can cause more problems than it solves. The second angle was the trend of exporting toxic wastes from developed world to developing world."

The opposition to Philadelphia's ash was bolstered by an October, 1987, report by the inspector general's office of the federal Environmental Protection Agency. It found that the regional EPA office had seriously underestimated the levels of dioxin in the city's ash.

Those findings have alarmed residents of the blue-collar Philadelphia neighborhood of Roxborough on the banks of the Schuylkill River. There, next to a city incinerator, a pile of 250,000 tons of stockpiled ash has grown to a four-story mountain of residue ringed by rusted auto parts. Residents now are afraid that pollution from the ash pile will leak into the Schuylkill and contaminate drinking water.

PART IV

INCINERATION:
PROMISE OR
PROBLEM?

CHAPTER 6

TRYING A EUROPEAN IMPORT

By Thomas J. Maier

In 1975, the city officials of Saugus, Massachusetts, found an alternative to a garbage dump that was polluting local waterways. Working with a new company, they opened one of the first American incinerators to use technology borrowed from Europe.

In the next four years, the plant would suffer repeated shutdowns, would require repairs that cost an added $11 million, and would need a financial bailout from the federal government.

In 1983, the company opened its second incinerator—a $160 million project to burn garbage and generate electricity in Pinellas County, Florida. It, too, would suffer mechanical problems and would require expensive repairs.

In 1984, the company, which is now called Wheelabrator Environmental Systems, opened a plant on the Hudson River in Peekskill, New York. Its mechanical problems have been less severe, but it, too, has proven to be unexpectedly costly. Westchester taxpayers have had to pay nearly $30 million since 1984 to cover a gap in electric revenues. Even Wheelabrator now concedes that

81

European garbage technology cannot be easily transferred to America.

A six-month investigation conducted in 1987 found that the problems in Saugus, Pinellas County, and Peekskill exemplify the multibillion-dollar risk that scores of towns and cities are taking on European incinerator technology as an answer to America's garbage problem.

Despite two decades of failures and cost overruns with all forms of incinerators, America is spending more than $17 billion on garbage-burning plants—an investment that experts predict will drive up the cost of garbage disposal for many households and could result in major mechanical and environmental problems.

Poised to profit from the huge investment is a largely inexperienced industry that has yet to demonstrate a long-term record of success with the technology it is importing from Europe.

By 1992, the study found, 115 new plants are scheduled to start up across the country—equaling the total number built in the past two decades. More than two-thirds of the investment will be in European-style plants, with average fees at least 24 percent higher than other kinds of technology. The typical price tag for an incinerator built by most major vendors exceeds $200 million.

And the taxpayers are paying for it—in Hempstead Town, where a $370 million plant is under construction, taxes for garbage disposal are expected to jump 137 percent by 1990.

Lee Koppelman, director of the Long Island Regional Planning Board, predicts that the $230 a year the average Long Island family spends to get rid of its garbage will soar. "Right now, garbage is out of sight and out of mind, with the town picking up the garbage at the curbside for a modest cost," he said. "But when you have to start paying the debt on these incinerators—with more and more stringent pollution control standards—we'll discover a sizable cost. I can see the cost paralleling the Southwest Sewer District, with people paying $700 or $800 a year for garbage disposal."

Incinerator builders are selling the plants to local officials in Hempstead and elsewhere as a proven technology that has solved Europe's garbage disposal problems. Some try to entice municipal officials to buy the plants with free trips to Europe to tour incinerators. But American plant operators are using the technology in fundamentally different ways than abroad—differences that already have caused costly problems at some plants.

"The only thing that is proven is that it's very costly and still has a lot of bugs in the technology," said Ken Woodruff, former chairman of the American Society of Mechanical Engineers' garbage division. "The average American official goes to Europe and sees the garbage going in, but they don't ask about the guts of the plants."

European incinerator technology, unlike its American counterpart, does not involve the sorting of garbage before it is burned, and produces steam rather than electricity. European plants are generally smaller, and the type of garbage burned is different. American garbage produces substances that cause more corrosion, particularly at the high temperatures needed to generate electricity.

In a report issued in the fall of 1987, Moody's Investors Service said that European-style incineration is not yet a "proven" technology in the United States, and warned that unscheduled shutdowns for all types of plants appear to be increasing. Moody's said: "A resource recovery plant project may not be suitable for every municipality, and such a project clearly entails major risks."

"We're still in the infancy stage with this industry—there are only a handful of plants nationally with any kind of operating history," said Edward Kerman, Moody's vice president. "I think it remains to be seen if it can work here. It has worked in Japan and Europe. But we have a different type of garbage here, and we have different environmental rules. There is no long-term history here."

Experts warn that local officials—with little help from state and federal governments—are rushing into the risky business of garbage incineration without fully grasping the potential problems. "They aren't looking sufficiently at recycling efforts," said Howard Levinson, who oversees the municipal waste program for the U.S. Office of Technology Assessment. "Everyone seems to be going whole hog for incineration."

WEST GERMANY COMBINES RECYCLING AND BURNING

By Adrian Peracchio

On a drab dead-end street in a gray suburb of Düsseldorf, West Germany, a densely populated industrial city, sits one of the answers to Germany's ever-growing mountain of garbage: a massive incinerating plant that burns more than 385,000 tons of trash a year.

A few streets away, in a shopping district like thousands of others throughout West Germany, is part of the other solution to this nation's garbage quandary: a housewife carrying a string bag bulging with unpackaged tomatoes, onions, and loaves of bread.

The shopping, cooking, and housekeeping habits of the German housewife at one end and the incinerating plant at the other are the alpha and omega of the way West Germany copes with its garbage. Although each has its own merits, it is together that they are effective. Incineration technology, German experts say, has been successful because it has been coupled with ingrained national habits—a willingness to hold down packaging and recycle waste materials.

But West Germans have by no means solved all the problems of garbage incineration. Ash disposal is still a major thorn; so is the problem of eliminating all dioxin emissions from flue gases.

And another problem has cropped up in a system that appears to enjoy a wide political consensus. It is a resistance movement similar to one occurring in the United States—growing opposition by environmentalists. The difference is that, in the United States, the opposition is led by one-issue civic groups. In West Germany, it is being promulgated by a rising political force—the environmentally conscious Green Party, which has given garbage high priority on its national agenda, just below nuclear disarmament and the death of forests. The party opposes the construction of new garbage incinerators and would like to see existing ones phased out.

Like their counterparts in most parts of Europe, German shoppers have been using string bags to do their marketing for as long as anyone can remember. The bag, and the way in which groceries are retailed in European markets—with a minimum of packaging by American standards—help a German produce, on the average, half as much garbage each day as an American shopper accustomed to discarding plastic wrappers and shopping bags.

Separate the German housewife and her habits from the incinerating plants, the experts say, and the nation's garbage-disposal strategy falls apart.

The habits include source-separation and recycling. Düsseldorf residents, like other West Germans, deposit their old newspapers and empty glass bottles in separate collection bins centrally placed in each neighborhood. In some cities, like Heidelberg, each family is required to separate organic refuse from recyclable materials and incinerator-bound waste in three garbage cans. In other communities, families are charged for trash removal according to the amount of garbage they generate—the less trash left out for pickup, the less they pay each month.

The incinerators at the other end of the system are exemplified by the Düsseldorf plant, which has been burning the garbage produced by more than 770,000 residents for two decades without a hitch. The large square building is almost clinically neat: There is no sign of garbage anywhere near it or even on the immediate premises outside the processing building; there is no smell and little noise.

For virtually every large West German city, incineration has been an accepted practice ever since the postwar industrial boom resulted in an increasingly affluent population overtaxing the capacity of dumps and landfills. "We ran out of room for garbage more than 20 years ago," said Karl Heinz Thoemen, who has managed the Düsseldorf incinerator since it opened. "Düsseldorf has been through every garbage problem that Long Island and New York are facing right now, and we had to devise our own solutions and make them work."

The problems of the New York metropolitan area are not unfamiliar to Thoemen, who has been consulted by Grumman experts working on incinerator design and by the builders of the first Hempstead plant. "The garbage problems here and in New York are not so different, so far as the specifications and requirements and the technology that's needed," he said. "But the political situation is very different. And people's habits are different."

Some of the differences are startlingly basic. West Germany, a nation of 62 million, has approximately one quarter the total population of the United States squeezed into a space the size of Oregon. Yet it produces only about two and a half times the municipal garbage generated each year by

New York City, with a population of about 7.1 million.

Americans generate too much garbage per person for the German solutions to waste disposal to work effectively, experts say. "Americans are producing much, much more waste than we are," said Professor Hans J. Karpe, director of the Institute for Environmental Protection at the University of Dortmund. "The waste strategy West Germany has adopted can be summarized simply: reduce, prevent, and burn. You cannot use just one of the components of this strategy to the exclusion of the others."

Since World War II, an increasing number of German cities have relied on sophisticated technology, state-of-the-art anti-pollution systems, and the frequent coupling of plant operation to the production of steam. The number of incinerating plants in West Germany has risen from 7 in 1965 to 47 plants in 1987 that burn more than 8.8 million tons of trash a year—about 35 percent of all trash generated in the nation. German authorities expect 20 more plants to be in operation by 1995, accounting for 50 percent of all waste disposal. They expect most of the remaining 50 percent to be recycled or landfilled.

Virtually all the plants use technology similar to that used or planned for major American plants. Perhaps the most significant difference is size. Even the larger German plants are about half the capacity of those planned in the New York area. Another major difference is that nearly 28 percent of all incinerated German waste is turned into steam for heating and electrical energy, compared to about two percent in the United States.

At the same time, dependence on landfills is being drastically scaled down. The state of Bavaria has gone from 4,000 to 40 landfills in the last 20 years and plans to eliminate most of those in the next five to ten years. Fears about the leaching of toxic materials and tighter air pollution standards have convinced most state and federal officials in Germany that incineration provides the safer alternative.

The German experience in waste disposal has provided a technological and social model for several other western European countries where limited space is a crucial factor. Switzerland, Sweden, the Netherlands, and Italy have taken West Germany's lead in turning to incineration. European nations also are taking a cue from the German system of combining incineration with recycling. And the German technology is being used in Japan.

But West Germany is exporting more than its technology. Some of the ash from its incinerators is "exported" to East Germany, which—unburdened by pollution controls and eager for foreign currency exchange—accepts it and landfills it for an undisclosed fee. And some of it has ended up outside national boundaries at the bottom of the North Sea—a practice that West Germany has recently agreed to stop.

And, in recent years, the German incineration program has triggered increasing resistance by environmentalists—led by the Green Party.

The Green Party is no less opposed to landfills. In their place, the Greens would substitute a strictly enforced nationwide program to recycle and compost all garbage. Each household would be required to compost all organic material and turn in recyclables to central collection points. Anything that could not be composted or recycled would be stored in large, leak-proof warehouses built on pilings so that no toxic material could leach into the ground.

"We calculate that 50 to 60 percent of all material that ends up as municipal waste should not be thrown away at all," said Harry Kunz, a 26-year-old economist in Cologne who is the Greens' national spokesman on waste issues.

Most waste experts and state and federal officials in Germany have dismissed the Greens' schemes as ideals that would be hard to enforce and costly to administer. But they credit the party with motivating local residents to support more ambitious recycling plans. More important, the Greens have been able to galvanize local residents into resisting the approval of new incinerator plants in their towns and blocking other plants from enlarging their capacity.

At Iserlohn, an industrial town about 40 miles northeast of Cologne, a 20-year-old incinerating plant was shut down in August, 1987, after a malfunction of its filtering system let out a cloud of dust that coated nearby streets. The malfunction was easily fixed, and plant operation returned to normal. But the county's waste management director, Olaf Moschner, who supervises the plant, said the public outcry is making it much harder for the plant to get permission to expand its capacity by renovating two of its three burners.

"We paid more than 200,000 marks [$125,000] for independent tests to prove that dioxin levels from the dust emission were negligible, too low to be a worry, but the people are still too worried," said Harald Schmidt, the plant's manager.

Among those who remain unconvinced in Iserlohn is Marlis Griesbach, a 35-year-old high school biology teacher who heads the local offshoot of the Environmental Protection League, a nonpolitical grass-roots group. "We want a better explanation of dioxins from that plant," she said. "The official position is that test

Control center at a Düsseldorf, West Germany, incinerator.

results show no danger because they are below the limits for human exposure, but those limits are politically determined. We always have to live with a certain amount of risk, but what's important is to limit those risks as much as possible."

Permissible dioxin levels from waste incinerators are still being studied by the German federal government, but waste experts generally agree that German pollution standards are among the toughest in the world for other emissions. Monitoring practices are similarly strict. If a plant exceeds permissible levels of any one of 12 monitored pollutants for more than 60 minutes, federal law demands the plant be shut down until the problem is corrected. Emission levels are reported every midnight to state environmental agencies, and a new system is being tested to link every incinerator to the state agencies by computer for instant monitoring.

Training requirements for plant workers are strict. West German incinerator operators receive six months of classroom study in combustion efficiency and dioxin formation and reduction. In other European countries, only upper management receives such training.

For all the safeguards, not all German environmental experts are satisfied. "It will not be possible for Germany to go without incinerators, because they are really necessary," said Helmut Schreiber, a research fellow at the Institute for European Environmental Policy in Bonn, who is critical of his country's incinerator program. "But I'm a bit worried about building too many of them. They are too expensive to build, not profitable to run for many years, and there needs to be a certain amount of waste generated so the plants can run efficiently. I would caution Americans to take a good look at them. They are not the only answer. Americans would also have to change their behavior to reduce the amount of garbage they produce. And they would have to be educated in how to do it. Public attitudes and the willingness to change habits are as important as new technology."

The industry is convinced the new European incinerators will work as expected. "From a technological basis, it's like saying that a Japanese car can't run here in America," said Robert Marrone, vice president for Pittsburgh-based Dravo, a major builder. "Of course it will. I think it's a proven technology and we'd enjoy proving that, and our management is convinced it'll work. That's why we're investing in it."

Others argue that, even if the risks are enormous, the alternatives are worse. "The state of the art for incinerators is very uncertain," said Islip Town Supervisor Frank Jones, whose town has opened the first of the new incinerators on Long Island. "Since we don't have landfills, and we know from the barge that shipping is not the answer, it seems that incineration is part of the answer," Jones said.

An examination of garbage incineration in America shows that:

- The European-style technology known as mass burn—which is being used at only 29 percent of existing incinerators—accounts for more than two-thirds of the new facilities being built. These new plants, on average, are nearly double the cost of all types of existing incinerators—with many exceeding $200 million. Half of the 32 mass-burn plants already working have experienced unscheduled shutdowns, and three others have been closed permanently.

- Plants based on American technology have suffered even more problems. Thirty-two of the 60 in operation have suffered unscheduled shutdowns of a week or more. Sixteen others have been shut down, well before the end of their expected 20-year lifespans. In the 1980s, incinerators have suffered expensive repairs, fires, and several explosions—including a blast at an Ohio plant that killed three workers.

- Few of the major American companies building the new plants have experience with the technology, and many are hiring Europeans for their expertise. Four of the industry's ten leading firms have never built an incineration plant of any kind; two others have built only one each. Those with no operating experience include the builders of the Hempstead, Huntington, and Islip incinerators. "It's on-the-job training for so many of these vendors—it's alarming to me," said William Robinson, an engineer and author of a book on incinerator technology.

- Electric utility customers are paying half the cost of most large incinerators—even if their garbage is not being burned at these facilities. Under current law, utilities must pay a premium for the relatively small amount of electricity produced by incinerators. On Long Island and in New York City, that rate is nearly double the utility's own cost to produce power. Long Island Lighting Co. Chairman William J. Catacosinos, whose company is suing over the price it must pay for electricity to be produced at Hempstead's garbage plant, said, "We on Long Island are subsi-

dizing these plants. . . . The community will be effectively paying twice what it pays to LILCO on the price of electricity."
- Despite the industry's claims of a proven track record in other
  countries, most foreign plants are no more than half the size of
  American incinerators, and more than 70 percent recover primarily steam or no energy at all. Hempstead Town's plant will be
  three times the average size of the 125 plants using the same
  technology abroad. And only six of the foreign plants produce
  electricity on a scale comparable to that required in Hempstead.
  To get the electric revenues guaranteed by U.S. law, incinerator
  builders must enlarge the typical European plants and add equipment that allows the burning garbage to produce electric power.
  That equipment not only boosts plant costs but also requires
  burning at higher temperatures and pressures. Experts say that
  this has already caused corrosion and other problems that have
  led to multimillion-dollar repairs.
- Taxpayers are increasingly bearing the financial risks of incinerators—through increased garbage fees and town taxes as well as
  electric rates—rather than the firms that are building them.
  Wary of the risks, Wall Street investment houses are increasingly
  demanding that cities and towns that want money for incinerators back the bonds with tax money. If a plant fails, the costs
  would be passed on to residents.
- Officials have largely ignored the problem of what to do when an
  incinerator shuts down for repairs. Many Long Island towns,
  with landfills closed or due to shut down, would face huge costs
  to transport garbage. A complete shutdown at the Hempstead
  plant now under construction would force the town to pay for at
  least 100 trips by trailer trucks each day to haul garbage off the
  Island, according to one estimate. A Westchester County study
  found that a shutdown of more than a month at the Peekskill
  incinerator could cost more than $5 million—forcing the county
  to declare a health emergency and ask for state aid.
- Recycling is undermined by contracts negotiated by the incinerator builders. They normally require that all garbage from a
  community be funneled to the plant, where such energy-rich
  recyclables as paper can be burned to produce electric power. "It's
  essential to the financing of these plants that they have a steady
  waste flow so the plant can be paid off," said Frank McManus,
  editor of the industry newsletter, *Resource Recovery Reports*.
  Despite a state goal of 40 percent recycling by 1997, most Long
  Island garbage is committed to be burned at the new incinerators.

"About seven percent of the nation's garbage was incinerated in 1987. This compares to less than five percent in 1984," said Robert Hunt, vice president of Franklin Associates, a Kansas consulting firm specializing in solid waste. "The proportion is expected to double to 14 percent by 1995," Hunt said. "In New York State, nearly ten percent of all garbage is now burned—about 1.8 million tons annually. The percentage could more than double by 1992 as new garbage-to-energy plants capable of burning more than 2.6 million tons a year come on line."

In some ways, the story of the incinerator industry can be told through the story of its most experienced company—Wheelabrator, the builder of the Saugus and Pinellas plants. The company now says there is a learning curve for U.S. vendors in transplanting European methods. And the company says this was learned only after long and costly problems at the two plants.

In company documents, Wheelabrator acknowledges that the European technology is not directly transferable to the U.S. for a number of reasons. Among them: smaller European boiler sizes; chemical differences between European waste and American garbage, which have caused mechanical problems; and steam temperatures and pressures in Europe that are generally lower than in the United States. "The other vendors are saying that they can transfer the European technology, but they have yet to do it," said Ronald Broglio, Wheelabrator's vice president of operations.

"There were very significant problems with Saugus," recalled company spokesman Kevin Stickney. "It was a European plant burning American waste. And within two to three years, we had to replace the superheaters, redesign the furnace, and redesign the combustion-air system—all toward accommodating this more abrasive and more volatile waste."

At the time, the federal government had a keen interest in the Saugus plant, along with several other projects using garbage to generate energy. During the 1970s, the Arab oil crunch and the nation's swelling landfill crisis made the government take a new look at incineration as a viable disposal method. The industry responded with a slogan: there is gold in garbage.

Saugus became appealing because it relied on the European-style process called mass burn, which was considered the least complex system. Garbage was tossed into a large furnace containing a boiler that heated water into steam, which was then sold to a nearby General Electric factory. It was used for heating and to operate machinery. Despite its much larger size, Saugus was comparable to most European plants, which produce steam rather than electricity.

Meanwhile, other technologies foundered, suffering many more problems than Saugus. Several projects, including the first Hempstead incinerator, which eventually failed, relied on an American-developed refuse-derived fuel (RDF) technology. Most RDF systems mechanically sorted out the garbage before it was burned and chopped it into fuel pellets that were then incinerated to produce electricity. Despite a financial subsidy from the government, the RDF technology proved to be too complex and troublesome, sometimes causing fires and explosions. Overall, 10 of the 19 permanently shut plants used the RDF technology.

Other processes also experienced problems. Several incinerators using a technology called pyrolysis—which extracted air from the garbage before it was burned—quickly failed. Modular incinerators, which are smaller prefabricated versions of the mass-burn plants, were crippled by chronic breakdowns.

During the 1980s, European-style incinerators became increasingly popular, as local officials searched for a proven garbage technology. As McManus, the *Resource Recovery Reports* editor, explained, "People don't want to be on the cutting edge of garbage technology. They want a system that will get rid of the garbage."

Larger and more expensive incinerators became viable with the passage of the 1978 Public Utilities Regulatory Policies Act, which required utilities to buy electricity from the resource recovery plants at the price the utility ordinarily would pay to produce the power. Some states, including New York, went a step further—setting fixed rates that were even higher. With electric sales defraying the cost of new incinerators, municipalities became eager to sign up. As a result, many new incinerators now added electric turbines, increasing capital costs by about 20 percent. And they also increased the temperature and pressure of their boiler systems to generate electric power.

One of the first big electricity-producing incinerators was the Pinellas plant, near St. Petersburg, Florida. In 1980, the company invited local officials—who had traveled to France, Germany,

Newsday

How It Works

The two major ways of incinerating garbage to produce energy

Mass burning process

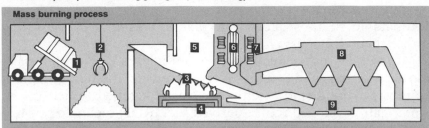

Preparation

1. Trucks are unloaded into an enclosed storage pit, large enough to store up to five days' worth of trash.

2. Plant operators use an overhead bridge crane to dump trash down hoppers to the furnace, selecting, to the extent possible, different types of trash — paper, wet garbage, etc. — from the pit to keep the fire burning the way they want it to.

Burning

3. Inside the furnace, the refuse falls on a grate above the flames.

4. The grate moves around to tumble and mix the garbage to allow even burning. The garbage is burned for one second at temperatures of at least 2,000 degrees F.

5. Forced air is injected above the fire to maintain temperatures of at least 1,500 degrees for 15 seconds to destroy most of the dangerous gases, such as dioxin, that develop from the burning of the garbage.

Energy production

6. The burning trash heats water in the boiler to generate steam. Hot air from the incineration is funneled into the boiler to provide additional heat.

7. The steam drives a turbine to generate electricity that is sold to a utility, or, less often, the steam is delivered to customers that use it to heat and cool buildings.

Aftermath

8. Gases left from the burning of the trash are blown through the boiler by a fan and treated by an air pollution control device.

9. The ashes (generally about 10 percent of the volume, and up to 35 percent of the weight, of the original garbage) are removed from the furnace by an ash extractor and cooled. Magnets pull metal residue from the ash for recycling. Typically, ash is disposed of in a landfill.

Refuse-derived fuel (RDF) process

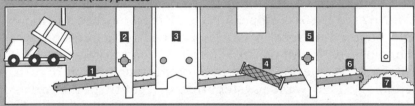

Preparation

1. Trash is piled one foot deep onto a conveyor belt.

2. The belt carries the garbage into a shredding machine where swinging hammers break open bags, smash glass, and loosen and expose refuse.

Sorting

3. Metals are removed from the waste by magnets and set aside for sale to a scrap dealer.

4. The remaining trash passes through a series of screens that sort out sand, dirt, glass, rocks and other noncombustible material for disposal in a landfill.

Preparing RDF

5. The remaining trash — about half of what was piled on the conveyor belt — goes through another shredding machine where it is smashed into small pieces.

6. This material can be sold to a utility as is (in which case it's referred to as "fluff" RDF). Or, it can be compressed into pellets or briquets (called "densified" RDF) before being sold.

Producing electricity

7. The utilities burn RDF, sometimes in conjunction with coal, to produce electricity.

PRINCIPAL SOURCES: National League of Cities, Combustion Engineering Inc.

and Switzerland to see the European technology in action—to a lunch of London broil Bordelaise, strawberry champagne floats, and chocolate mousse at the ground breaking.

Trouble started for the $160 million Pinellas plant soon after it opened in 1983, as the difficulty in producing electricity from American garbage became apparent.

OPPONENTS TURN UP THE HEAT

By Irene Virag

Armed with names like Mothers Opposed to Mass Incineration, or MOMI, Stop Incineration Now, or SIN, and Citizens Reacting Against Pollution, or CRAP, new civic groups are coming from the grass roots to wage the protest of the '80s: the community crusade against garbage plants.

Propelled by its passion, the anti-incineration movement, like the anti-nuclear movement, is a growing political force. "What's happening here will affect the next town supervisor's election," predicted Paul Thurman, an officer of the Coalition to Save Hempstead Harbor, which is fighting a proposed $219 million, 30-story incinerator project that would change the landscape of a century-old Port Washington sandpit and—as far as its opponents are concerned—threaten the health of residents by polluting the surrounding area.

"And it will go beyond that. Garbage management is an issue throughout the country," Thurman said. "We don't want any incinerator in any community anywhere."

In Delran, New Jersey, mothers with baby carriages blocked trucks as they arrived at a mass-burn facility; in Preston, Connecticut, Mohegan Indians in ceremonial headgear and beads chanted prayers to protest plans to build a plant on sacred burial grounds. In Minneapolis, members of Greenpeace chained themselves to a fence at the construction site of a $140 million incinerator. The residents of Montgomery, Maryland, wore "Ban the Burn" buttons in opposition to a proposed plant.

In Blackshear, Georgia, incinerator opponents carrying "Dump-Busters" signs gathered 3,000 signatures on a petition to defeat a trash-burning plant. In Pennsauken, New Jersey, 60 members of the group called SIN carried lit candles and cardboard signs that read "Resource Recovery, My Ash" as they paraded through the audience at an evening outdoor concert. Thousands of Hasidic Jews—wearing traditional long black coats and felt hats—marched across the Brooklyn Bridge in 1984 to protest a proposed 3,000-ton-a-day incinerator in the Brooklyn Navy Yard that is expected to cost $445 million. "We are determined to fight this to the end," said Rabbi Chaim Stauber of the United Jewish Organizations of Williamsburg.

Although the new issue incorporates the not-in-my-backyard mentality endemic to traditional suburban zoning battles, it has taken on an added dimension—a moralistic approach that wraps the old motherhood-and-apple-pie message into a contemporary package. "To call us NIMBYs only hardens our resolve. This is an ethical, moral community," Thurman said. "To accuse us of such limited motives is an insult. We're the baby boomers. . . . We've worked hard—now we want to put something back into the community. We're educated and skillful— we're a generation of good communicators."

The industry has a different view. "The public should have a say in the management of its garbage," said David Sussman, a former Environmental Protection Agency official who is a vice president of Ogden Martin Systems, one of the nation's leading incinerator builders. "However, the small groups of anti-everything and outside environmental evangelists who come in with the express purpose of preventing a community from dealing with its garbage problem are bad. . . . They don't help local communities dealing with a serious social issue like garbage." Town officials such as Islip Supervisor Frank Jones concede that the anti-incinerator groups have been able to influence decisions because "nobody wants to stand in public and argue the other side." But he added that, when the public realizes the extent and cost of the garbage crisis,

it will be easier for officials "to make the tough decisions on siting a landfill or ashfill or resource recovery plant."

The anti-incinerator ranks have been raising a stink over the last decade—following up fights to close landfills with struggles to stop the garbage-burning plants that have replaced them. They are not—as they make quite clear— just protecting their property. They say they are saving the harbor, cleaning the air, preserving the environment, safeguarding the health of their children. "This issue touches the family," said Louise Tripoli, a founder of the Coalition to Save Hempstead Harbor—a group dedicated to cleaning up the harbor, where seven hazardous waste sites have been identified. "How can I be in favor of anything that is a threat to my health and the health of my husband and children?" Incinerator advocates and some government officials say the health threat posed by the plants is minimal.

Often, the incineration issue spreads across local boundaries. It is impossible to put a fence across the sky or hold back the wind. On Hempstead Harbor, the southerly winds in the summer might carry emissions across the water—stirring concern in the communities where the coalition is active. And in Los Angeles, inner-city and suburban community groups forged an alliance that defeated a controversial proposal to build three huge incinerators. The campaign underscored the politicalization of garbage.

When the first plant was proposed, officials and the builder, Ogden Martin, hailed it as an up-to-date answer to the city's garbage problems. But residents felt otherwise. Schoolteachers, mothers who do not work outside the home, and ministers—many of whom had been active in anti-crime campaigns—learned the language of incineration and spoke at public hearings. They asked questions about potentially toxic ash and the estimated 220 garbage trucks that would rumble through their neighborhoods every day. They lobbied the city council for an expanded review of the $235 million facility that would have burned at least 1,600 tons of garbage a day—about one-tenth of the city's total daily waste pickup. They enlisted the support of mayors and council members and homeowners of nearby cities—who feared that emissions from the inner-city incinerator would blend into the smog. Diverse community groups formed a coalition—inner-city blacks from Concerned Citizens of South Central Los Angeles banded together with yuppie lawyers from a suburban group called Not Yet New York.

In local elections in June, 1987, the battle to stop the incinerator was credited with the defeat of the city council president and a candidate supported by Mayor Tom Bradley, who had backed the plant. Soon afterwards, Bradley trashed the incinerator plan.

Some skirmishes have been lost. In Camden County, New Jersey, community opposition could not prevent the approval of permits to break ground on an incinerator. But the war against incineration continues. On Long Island, the battles against garbage disposal have raged in Plainview and Port Washington, in Old Bethpage and Oceanside, in Islip and East Northport, in Riverhead and Westbury and Wyandanch.

In the battle over the North Hempstead incinerator, the community's strategy has encompassed the colorful, the confrontational, the civilized.

The poster that heralded the group's entry into the incinerator war screamed in bold black letters: "WARNING. A 1.8 million lbs. per day mass burn incinerator is to be built in our community. We can stop it if we act NOW!" The photocopied poster envisioned dangers in the emissions from such plants—"thirty polychlorinated dibenzo-p-dioxins, forty dibenzofurans, and more than forty other polynuclear aromatic compounds."

In November, 1987, the Coalition to Save Hempstead Harbor hosted a Gatsby-style cocktail party to raise funds for its legal challenge. In the elegant rooms of a mansion overlooking Roslyn Harbor, plastic surgeons chatted with music promoters, architects mingled with environmentalists. Men in red ties nibbled on chocolate-covered orange rind as they wrote checks for hundreds of dollars. Earlier in the battle, some of the same people—as well as members of other opposition groups— had testified at public hearings and picketed town hall. They sponsored a "Save the Harbor Sail" that brought 2,000 protesters to a Sea Cliff beach. They staged a quiet candlelight vigil outside the home of North Hempstead Supervisor John Kiernan and a more vocal, impromptu demonstration that ended in threats to incinerate Kiernan's residence.

The coalition and other area citizens' groups united in legal attempts to block the town's moves to acquire the site, obtain construction permits, and select a vendor. Proponents of the plant recognize the community opposition as a potent force. "I can't condemn community groups for being concerned about a project of this magnitude," said Bert Cunningham, executive assistant to Kiernan, "but the opponents to this plant haven't been open to rational discussion. Their response

Boiler tubes and the superheater that push steam through the plant's electrical turbines began to erode and break. The high concentration of plastics in the garbage, combined with the high temperatures, caused acid gases to build up in the Pinellas boiler. Repairs cost $8 million. The unscheduled shutdowns caused a buildup of 25,000 tons of garbage, which cost another $780,000 to cart away. Overall, Wheelabrator failed to produce enough electricity to meet its guarantee for 9 of the first 20 months of operation. The company says it paid for the losses. But in 1985 alone, the price of dumping garbage at the plant jumped 36 percent.

Wheelabrator's problems have not discouraged other firms from jumping into the resource recovery business. Some major vendors—like American REF-FUEL and Waste Management Inc.—come primarily from the waste-hauling business. Others, like Combustion Engineering, Dravo, and Ogden Martin, are engineering firms looking to enter a lucrative new market. "Some of the nuclear construction firms are looking for a new line of business, now that that field is dying out," said Woodruff, of the

was a simple 'No. We're against it. Stop it.' Their legal challenges could add delays. They might achieve a compromise modification—like some kind of change to the xyz monitoring device in a computer panel somewhere or a reduction in the size of the plant. But they aren't going to achieve the ultimate success of stopping the plant."

As the fight continues, the coalition is calling in technical experts to formulate a large-scale recycling effort—the alternative of choice among incinerator opponents.

Not every opposition group uses the same strategy. Brian Lipsett of the Citizens Clearing House for Hazardous Waste in Arlington, Virginia, favors a broad base of support in the community. "Successful fights are based on action, not words," said Lipsett, whose group counsels grass-roots organizations. His weapons of choice include pickets and protests and signs and balloons. And a war plan that is 90 percent political, 5 percent technical, and 5 percent legal.

"The words of lawyers and engineers aren't going to win this battle," Lipsett said. "If you're able to raise enough money to hire one expert, you can be sure that they have ten experts to say the opposite."

American Society of Mechanical Engineers. And they have latched on to the big mass-burn plants.

Several experts say that municipal officials are ill-equipped to sort through the complex technologies offered by resource recovery firms and are often swayed by misleading claims. "There is a certain amount of hype connected to these firms, who say they are going to make sure that these plants are going to work," said Jeffrey Peirce, a professor at the Waste-To-Energy Production Lab at Duke University. "The people who are building mass-burn plants are making the same sort of claims as those who built plants in the early 1970s which failed."

Many communities may end up locked into incineration, because contracts require all garbage to be burned. The industry calls it flow control guarantees; environmentalists say it kills any hope of recycling.

Ultimately, the risks of incineration come down to cost—both to the pocketbook as well as to the environment.

Some firms say they are assuming the financial risks, with enough equity to back up any new projects, financial statements

Workers fix the grates in the boiler section of the North Andover, Massachusetts, garbage-burning plant.

show. Other firms, to establish themselves in the market, have begun to overextend their finances. Dravo Corp. suffered a $22 million loss for the second quarter of 1987 because of cost overruns at two of its incinerator projects. Dravo went into technical default when its net worth was listed at $171.8 million—its $68 million in loans require $190 million in net worth. "It's got to be viewed as a new business," said Dravo's Robert Marrone. "Even if you take a beating every once in a while, you still have the operating period later on in which these facilities will make money."

The industry says that Wall Street's willingness to finance new projects underlines its faith in the new garbage technology. But Wall Street analysts say their main concern is that financial guarantees for the bondholder be met—and not whether the incinerators will actually work. "We cover ourselves so that the

length of the bonds is not tied to the length of the project,"
Standard & Poor's Timothy Tattam said.

With changes in federal tax law and proposed changes in the
Public Utilities Regulatory Policies Act, the potential prof-
itability from incinerators has dimmed for some vendors. Wall
Street is asking for increased guarantees by taxpayers to back up
new projects. Said Moody's analyst Alfred Medioli, "When you
talk about risk, you have to ask, 'Who owns the plant?' And the
ownership role is more and more falling upon government."

Although investors and the private companies may be pro-
tected in the event of a plant failure, there is little protection for
the municipality served by the plant. Its garbage may pile up, and
its taxpayers will have to foot the bill for another way to get rid
of it.

Wheelabrator's Broglio said, "Basically, there are big companies
that are guaranteeing these plants. [Municipal officials] figure if
it doesn't work, they won't suffer. But that's not true. They will
suffer if the vendor walks away."

FATAL FLAW IN AN EARLY PLANT

By William Bunch

Kenny Bell's story reads like a war story: the sudden fireball, the screams of his fallen colleagues, the scars that have blackened his once-agile hands, and the nightmares that haunt his sleep.

Bell was not injured in combat, however, but in a garbage incinerator explosion. In December, 1984, he was working in the loading area of the Akron garbage-to-steam plant when an explosion—blamed on illegally dumped hazardous wastes—ripped through the facility, killing three plant workers and injuring seven others, including Bell.

"When I came into the control room, people didn't even know who I was," Bell recalled. "My plastic helmet . . . had melted and had burned a ring around my head. All the hair was burned off my body, my face, my uniform was burned up. One of the workers named Scotty helped put me out—he was knocking some of the fire off me with his gloves."

As Bell struggles to recall the accident, his voice drops to a near-whisper and he buries his head in his hands, while his wife, Tamara, caresses his shoulder.

The Akron plant, which opened in 1979, used an American-devel-oped technology that was popular at the time. Today, the majority of plants under construction use the European mass-burn technology. The plant has cost more than $76 million—including $20 million for repairs and improvements. The federal government has stepped in twice to bail it out with $32.7 million in grants. After the fatal blast, it reopened in late 1985 with a new contractor in charge. Officials now agree that the 900-ton-per-day plant is finally operating smoothly.

But industry officials also say the long string of problems at Akron in its first five years of operation dramatically points out both the difficulties of using untested technologies to get rid of garbage and the ongoing risk of explosions at other new plants.

"There's no way that you can eliminate explosions in the solid waste process," said Andrew Martin, the chairman of the solid waste division of the American Society of Mechanical Engineers. "Knowing that, you can design a facility to minimize explosion damage."

Unfortunately, the experts agree, the original design of the Akron plant did not do that. But there were few other garbage-burning plants in America as design models when the Akron plant was conceived in the mid-1970s. The nation was in the middle of an energy crisis, and the Akron plant was viewed chiefly as a low-cost way of supplying steam heat to 160 downtown customers, not as a means of garbage disposal. Also, as was the experience of other cities with similar plants using the refuse-derived fuel technology known as RDF, Akron officials found little market for materials such as the glass and metals that were recovered.

In Akron, the plant was further hindered by experimental design features that were modified in 1982 with help from a $19.7 million grant from the federal Environmental Protection Agency.

A Canadian firm, Tricil Resources Inc., reopened the plant in 1983. But the plant could not compete very well for garbage against local landfills because of low tipping fees. To keep the plant operating, Tricil attracted industrial waste from outside of Ohio, but experts say these wastes were prone to explosions.

The big blast, which took place December 20, 1984, was later linked to a shipment of sawdust mixed with oil and paint wastes from a Kearny, New Jersey, factory. The New Jersey firm was cleared in a criminal case. Akron attorney William Zavarello said lawsuits filed by the families of the three dead workers, Bell, and another injured man against the firm, Tricil, the city, and other parties were settled in the range of $1 million to 1.5 million each.

But Bell, 52, says no amount of money could make up for his injuries. A former military police officer who once stayed active with karate, motorcycling, and 100 push-ups a day, Bell has been unable to work since the accident and finds sunlight painful because of his burns. A ring from the plastic helmet he wore is still visible around his head. He says he often wakes up screaming with flashbacks of the accident.

Bell also said he is not surprised that the plant, now owned by the city and run by a Massachusetts firm, has reopened, with the help of a new $13 million federal grant. "It's big business," he said. "So a few lives get lost."

CHAPTER 7

THE DEBATE OVER DANGER

By Michelle Slatalla

From planning studies to actual tests, estimates of air pollution from garbage-burning plants in the United States vary radically and use conflicting methods, fueling a national debate over whether incinerators pose a danger.

As a result, predictions of how much air pollution will be emitted from incinerators planned for Long Island and New York City vary greatly, even though all the plants will use similar technology and pollution-control equipment.

But while predictions of private consultants hired by local officials differed by as much as 2,800 to 1 for the emission of certain pollutants, estimates were nearly identical in predicting very little cancer risk.

The Environmental Protection Agency has reported that the risk posed by incinerators appears to be small. But critics say the scope of these studies and the quality of the testing done at the nation's 93 operating incinerators are insufficient to produce reliable results.

Unlike countries such as West Germany—which requires

daily reporting of air-pollution rates from every incinerator—the United States does not mandate regular and uniform tests. In many cases, plant operators are allowed to burn garbage for years without tests—or to test their own facilities without government supervision.

"There's something very wrong with this system, where we don't really know which plants are best because we have to rely on very incomplete data," said Dr. Daniel Wartenberg, a professor of environmental medicine at Rutgers University in New Brunswick, New Jersey. "At the plants, they know when the tests are going to happen, they get the plant tuned up, they pick a day when it's operating under the best conditions. What does that tell you about how a plant operates over time? Nothing."

Since Hempstead Town's first incinerator was found to be emitting the highly toxic compound dioxin in 1980, the fear of air pollution has spurred community groups, environmentalists, and some scientists around the country to call for stricter controls on garbage-burning plants. In Massachusetts, state officials closed down an incinerator in Lawrence for five months in 1987, after tests showed it was emitting too much dioxin. Some members of Congress are calling for federal emissions limits for two dozen air pollutants—a step the EPA has been reluctant to take.

Officials responsible for monitoring the plants say the planning studies provide an adequate prediction of the "worst" scenario of what the plants might emit and that the tests are reliable. "It's a classic question we ask here: How much information is enough information to make a decision about safety?" said Arthur Fossa, who heads the New York State Department of Environmental Conservation's air division. "Where do you stop? How much is enough? We don't know everything, but we know enough to make a responsible decision. We know the plants are safe."

The local questions about air pollution are a microcosm of the uncertainties that plague scientists and public officials across America as municipalities plan to operate more than 200 incinerators by 1992. While the federal Environmental Protection Agency has estimated that the health threat from inhaling pollution will be minimal—from 4 to 60 excess cases of cancer nationwide are expected annually when the plants are built—the agency has not estimated how much the threat will increase from other methods of exposure, including absorbing pollution on skin and eating contaminated food.

In New York State, studies predicting the extent and threat of air pollution—estimates based on numbers private consultants gather—must be commissioned by every municipality that builds a garbage incinerator. But the *Newsday* study, conducted by Maryland-based biologist Bernd Franke, who has conducted waste management projects for the German Marshall Fund, the state of Missouri, and the Philadelphia City Council, found that the local studies were based on "arbitrary assumptions" and did not consider the impact of all methods of exposure to air pollution. In addition, the studies did not consider the cumulative effect of clustering 13 incinerators in the New York City–Long Island area.

Locally, pollution predictions vary widely. For instance, according to the local studies, the plant under construction in Islip Town will emit 17 times the arsenic of the planned North Hempstead plant and 13 times the hydrogen chloride of the incinerator being built in Babylon. The plant slated for Huntington is expected to emit 2,800 times the cobalt of the planned Brooklyn plant. But all the plants are expected to increase the cancer risk only minimally—by fewer than two excess cases of cancer per million people, a ratio widely considered the threshold for "acceptable" risk.

"Nobody really knows precisely how much pollution will be emitted or what the health threat will be," Franke said. "The towns were just guessing when they conducted their studies, and the ways they chose to report their data are very arbitrary."

But local officials and private consultants who work for the industry and municipalities say the emissions predictions vary because the spotty nature of incinerator testing in this country provides incomplete data about what to expect from the new plants.

In addition, "waste composition can differ from day to day, so the numbers that you see also reflect different estimations," said Ben Miller, the New York City Sanitation Department's director of public policy for the office of resource recovery. "The factors do vary greatly, but in the minute quantities that we expect will be emitted, the differences would have no real significance on the risk."

But Franke said the local studies did not indicate that emissions rates were based on waste composition.

Some experts concede that the studies are imprecise—but say they are the best that can be done. "They are of limited value, but they are all we have right now; it's a flaky area," said Nicholas Stevens, an engineer with the Interstate Sanitation Commission in New York City. "Our public demanded that we look into health risk before we build these plants, but the method we have to look is pretty crude. Maybe now, everyone's method will converge in 20 or 30 years and you won't see the differences."

Consultants commissioned to prepare the studies say they make conservative predictions about emissions in an attempt to ensure that the plants will not pose a greater danger than expected. "What one tries to do is estimate conservatively the upper limits of what a plant will emit," said Ian Thomson, a senior associate of CSI Resource Systems, which prepared studies for the town of Hempstead's second plant. "Most of the pollutants will have no significant effect on public health."

Using the predictions generated by the local studies, Franke found that municipalities may have underestimated the risk from dioxins and furans, pollutants responsible for much of the potential health threat. Although the studies done for local governments all predicted that dioxin and furan exposure would cause about one excess cancer case per million people over a 70-year period, Franke found that some areas, particularly Glen Cove, might face a greater risk because of the cumulative pollution from several plants. North Hempstead, Hempstead, and Oyster Bay are planning to build plants within 15 miles of the Glen Cove incinerator.

In that area of highest exposure, Franke found, the risk of getting cancer from inhalation of dioxins and furans might be 21 times higher than local studies predicted. The Glen Cove incinerator has never been tested for emissions of dioxins and furans. To reach the health-risk conclusion, Franke plugged in numbers the federal government estimates would be typical for a plant of Glen Cove's type. Miro Dvirka, an engineer who designed the Glen Cove plant, said he does not believe emissions from that plant are greater than emissions from others.

DEC official Fossa said that the state is now testing all operating plants for dioxin. "You have to look at each plant based on the

Garbage In, Emissions Out

Common pollutants produced by a typical state-of-the-art, mass-burn garbage incinerator. The U.S. Environmental Protection Agency has not set air-quality limits for most of the pollutants. Descriptions below include possible health effects from heavy or constant exposure. But the EPA and the incinerator industry argue that the tiny amount of pollutants emitted from a typical plant pose little health risk.

From the smokestack

Particulates
Solid matter in soot or smoke. Other pollutants adhere to particulates and are deposited on lawns, cars and kitchen counters in the vicinity of the incinerator.

Sulfur dioxide
A gas produced by burning organic materials such as leaves and vegetables. Contributes to acid rain and in large doses can cause respiratory problems.

Nitrogen oxides
Odorless gases produced during combustion of fossil fuels and organic materials. Contribute to acid rain and smog. Can cause irritation of eyes, mucous membranes and lungs.

Hydrogen chloride
A colorless, nonflammable gas produced when dyes, artificial silk and paint are burned. Contributes to acid rain. In large doses, can be highly corrosive to eyes, skin and mucous membranes.

Hydrogen fluoride
A colorless gas or liquid with a strong odor. Contributes to acid rain. Irritates eyes, nose and throat and in large doses can cause burns and lung problems.

Carbon monoxide
A colorless, odorless, tasteless gas produced when material containing carbon is burned. Can reduce the blood's ability to carry oxygen, which elevates the heart rate.

Arsenic
Arsenic compounds are used in numerous products, including insecticides, paints and ceramics. Exposure to large amounts of arsenic, or small amounts over a long period, can cause kidney damage and blood problems.

Cadmium
A potentially cancer-causing metal, usually electroplated onto metal compounds to protect them from corrosion or used in batteries.

Chromium
A metal used in chrome plating and copper stripping that can irritate the respiratory tract.

Lead
A metal that can harm the central nervous system, blood-forming tissues, kidneys, liver and gastro-intestinal system.

Mercury
A metal used in industry and, as a fungicide, in agriculture. Inhalation of mercury vapor or absorption of mercury through the skin can damage the kidneys and the central nervous system.

Dioxins and furans
Dioxins and furans are families of organic compounds. Cause cancer in lab animals; considered a probable cause of cancer in humans.

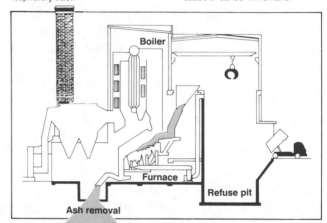

In the ash

The primary pollutants in ash are cadmium and lead, for which the EPA has set standards. (Scientists recently have started measuring levels of dioxins in ash and are still studying their significance.) Here is a sampling of the results of recent tests on mass-burn incinerators using EPA-approved tests.

EPA Limits:
Cadmium — 1.0 mg per liter
Lead — 5.0 mg per liter

Facility	Year of test	Cadmium	Lead
Baltimore	1985	0.2	0.6-2.4
Baltimore	1984	0.83	12.8
Baltimore	1987	0.4	8.4
Westchester	1984	0.3-1.0	0.5-3.7
Waukesha, Wis.	1985	1.33	11.3
Chicago	1983	0.71	5.8
Chicago	1981	0.25	0.34
Gallatin, Tenn.	1983	0.24	6.4
Hampton, Va.	1983	0.5	10.3
Philadelphia	1984	0.3	1.8
Nashville, Tenn.	1981	N/A	0.7
Saugus, Mass.	1981	0.8-5.3	6.7-31.0
Brooks, Ore.	1986	N/A	4.94-32.58
Baltimore	1986	N/A	0.79-3.02
Brooks, Ore.	1987	N/A	3.35-8.4
Glen Cove	1987	0.38-0.49	2.6-6.5
Westchester	1987	2.5-2.8	19-21

requirements at the time that the permit was granted," Fossa said. "In some cases, we may go back and ask for more information." Incinerators in New York State must pass a pollution test when they open, but the list of pollutants tested varies from plant to plant. And the frequency of follow-up tests also varies.

"You continually refine the best possible method, but you're always dealing with a margin of error in the estimates," said Mike Cahill, general counsel to the Islip Resource Recovery Agency. "When you get down to predicting a couple of years in advance what the particular emissions are going to be, it can get a little iffy."

But the state's system of having each town do its own narrow study of the possible environmental effects is flawed, some say. "It's insane for each town to decide for themselves," said Lee Koppelman, director of the Long Island Regional Planning Board. "They are not experts. It's too complex and expensive an issue to be done in a business-as-usual way."

Even after the plants are operating, some questions will remain because attempts to assess what is coming out of the smokestacks vary widely in quality, and states do not strictly enforce pollution limits. *Newsday*'s survey of 50 states, 5 territories, and the District of Columbia shows that at least 11 allow operators to conduct their own tests without government officials present. Some plants, like the Saugus, Massachusetts, incinerator, test annually for just a few pollutants. Others, like the Brooks, Oregon, plant, test for a broader range. At least one so-called state-of-the-art incinerator—in Baltimore—has been allowed to operate since 1984 without updated emissions tests.

The Baltimore incinerator, like all but about 20 of America's 93 operating plants, has never been tested for emissions of furans and dioxins, organic compounds that cause cancer in rodents and are considered a probable cause of cancer in humans. New York State has not set emissions limits for dioxins or furans.

The actual extent of the health threat will be based on numerous factors, including how efficiently the incinerators burn garbage. Dioxins are formed by colliding hydrogen, carbon, and oxygen atoms released by burning garbage. The hotter the burning chamber, the faster the atoms move and the less likely they are to bond into dioxins. But it is a trade-off. More pollutants that cause acid rain are created at higher temperatures. That is what happened at the incinerator in Brooks, Oregon, where state officials raised the plant's emissions limit for nitrogen oxides after tests showed the plant was exceeding its permit level.

The gases and tiny particles from the burning chamber are filtered through pollution control devices that capture all but the tiniest bits. And so the black, rolling waves of smoke that identified the stacks of old-time incinerators have been replaced by colorless, odorless particles that land on lawns, houses, and cars in a wide radius around the plant.

The studies commissioned by local officials considered only the pollution effects within a radius of less than 20 miles from a single plant. Franke concluded that pollution from the Long Island planned plants will be deposited all over the metropolitan area, in the ocean, and on neighboring states.

"But the overall cumulative impact is not that big," said Stevens of the Interstate Sanitation Commission. "With all this uncertainty, we do have a lot of different answers, and the numbers have a limited value. But we're learning."

"In the drug and chemical industries, the onus is on the companies to prove their products are safe," said Wartenberg, of Rutgers. "In the case of incineration, the onus is on the public to prove it's not safe."

In West Germany, home of much of the technology being used in the new American incinerators, air pollution monitoring is stricter. The law requires plants to report emissions rates for 12 pollutants each midnight, and a computer system is being developed to link all 47 incinerators to a continuous system to track their performance. If a plant exceeds emissions limits for more than an hour, it must be shut down. And if operators break the law, they go to jail.

CHAPTER 8

THE "TRASH" OF INCINERATION

By Michelle Slatalla

Thousands of tons of incinerator ash have been spread under Long Island roads and parking lots since the summer of 1986, despite state tests that show it often contains unsafe levels of toxic pollutants.

A contractor distributed up to 30,000 tons of ash from the Glen Cove incinerator to road builders, a practice environmentalists call "unbelievable" against the backdrop of a national and local debate over how badly ash contaminates drinking water. After a tour of the plant, state officials said they had no evidence that it was unsafe and, in any case, no authority to stop the practice.

Uncertainty over the toxicity of ash convinced the entrepreneur to halt sales. "Every time I'd see a truck come in, I'd feel like Al Capone," said Smithtown contractor Raymond Schleider. "I'd think, 'Am I going to go to jail today?' So even though the state didn't tell me to stop, I stopped."

No one in science or government is offering Raymond Schleider easy answers. The mound of gritty black particles now piling up outside the Glen Cove incinerator is a relatively small, but dramatic, example of the environmental risks that America is taking as the push to burn garbage accelerates. By 1992, more than 200 incinerators will generate about 55,000 tons of ash a day nationwide. For every 1,000 pounds of garbage burned, about 300 pounds of ash will be produced.

The nation's mounting ash pile serves as a constant reminder that incinerators cannot solve all of America's garbage problems. Dumps will still be necessary for ash—increasing the risk of groundwater pollution, as leaking ash creates a toxic soup whose potency has not yet been measured.

The planned Brooklyn Navy Yard incinerator—the first of five huge garbage-burning plants slated for the city—is expected to produce some 900 tons of ash daily. City officials expect to dump the ash at the giant Fresh Kills landfill, which has no protective liner, and some environmentalists believe the ash could add to New York's pollution woes.

The ash problem "is the last remaining monkey wrench standing in the way of plans to use incinerators as a cure-all," said Richard Denison, a scientist for the Environmental Defense Fund in Washington. "No matter what you do to improve incinerators, they are still going to produce ash. And we still have to decide how to handle it safely."

But as the debate continues in the United States, some incinerator ash is being used to build roads in West Germany, which also exports ash to be buried in East Germany.

Incinerator manufacturers say the danger of ash has been exaggerated. "Give me a bowl of ash and I'll eat it," said David Sussman, vice president of Ogden Martin Systems Inc., which is building an incinerator in Babylon.

Ash is of great concern on Long Island, where at least seven other incinerators are planned to operate by 1992. If all the plants operated at capacity, they would produce about 2,050 tons of ash a day, adding up to 750,000 tons a year—6.5 times the weight of Long Island's potato crop. But ash is a harvest that no one wants. Even as hundreds of millions of dollars are being invested in incinerators, their builders are having trouble finding places to dump the ash, and citizens' groups have sprung up to fight every proposed site.

Under pressure from the incinerator industry and municipalities around the country, the federal government has opted to exempt ash from consideration as hazardous waste, even though repeated tests show it often contains unsafe levels of pollutants.

"I'm not saying this is totally rational," said J. Winston Porter, administrator for solid and hazardous waste at the Environmental Protection Agency. "I'm just saying that if you read the law the way it's crafted, ash would be exempt." The agency argues that federal law exempts household garbage and that ash should fall into that category.

The EPA has drafted guidelines requiring that ash dumps be monitored for pollution. But there are no hard and fast rules governing the disposal—or use—of ash. When Philadelphia tried to ship its ash to Panama in 1987 for use in road building, the inspector general's office of the EPA issued a report saying that the practice would pose a serious health threat. Panama refused to take the ash.

From New York–area citizens' groups who passionately oppose plans for ash dumps, to Oregon residents concerned about ash tests that have proved inconclusive, the big question is: Just how dangerous is the chunky residue?

"Ash is full of junk. And it's not homogeneous; each sample is filled with different junk," said Michael Downs, administrator of the solid and hazardous waste divisions of Oregon's Department of Environmental Quality. "It's hard to decide what to do with it when we're not always sure what's in it."

There is no consensus among scientists to support the assumption that burying ash is safer than burying garbage. But a survey of state officials who oversee garbage regulation shows that they generally believe incinerators create less pollution than landfills, which produce methane gas and toxic leakage.

As the debate goes on, America's incinerators continue to produce the dense dark chunks of ash:

It begins in 100-foot-high trash mounds—resembling unwieldy gifts wrapped in wispy strips of paper—piled up in every incinerator.

Sharp, curved fingers of huge cranes attack each pile, grasp five tons of trash at one time, and drop their burden into a burning chamber, where it is engulfed by 40-foot-high flames so hot they color the air a pale yellow.

The burning garbage oozes molten down the chamber's chute, losing its shape and bulk and turning into black chunks of ash

MAYORS AND BIG BUSINESS: SIDE BY SIDE

By Marie Cocco

In January, 1987, the Philadelphia City Council was told that municipal garbage incinerators equipped with proper pollution-control equipment pose no "observable" threat to human health.

In May of that year, the Pennsylvania House Conservation Committee was told that placing a state moratorium on building garbage incinerators because little is known about their environmental hazards would be like putting a moratorium on automobiles—because little is known about how dioxin is formed in car engines.

In September of 1987, the Suffolk County Legislature was told that tests designed to assess the environmental hazards of incinerator ash—tests that often show the ash contains hazardous levels of pollutants—should be "regarded as not statistically valid."

These bits of controversial advice on how state and local governments should deal with one of the toughest garbage-disposal issues they confront—the environmental risks of incinerators—came from an arm of the U.S. Conference of Mayors. At first glance, it seems to be a case of one large and respected group of public officials—the mayors' conference—trying to help others.

But a closer look at the National Resource Recovery Association, the arm of the mayors' conference that lobbies on waste-to-energy issues, shows that it is partly funded by some of the businesses who have a stake in building incinerators.

Incinerator vendors, investment banking houses, consultants, and law firms that stand to profit from the construction of incinerators comprise about half the group's membership, association officials acknowledged. And the $225 annual dues paid by these private-sector members contribute to its budget. The rest comes from dues paid by cities who belong to the 200-member Resource Recovery Association.

"In general, what is beneficial to the cities is beneficial to the private-sector members," said Ronald Musselwhite, director of the association.

The group was formed in 1982, Musselwhite said, after cities asked the Conference of Mayors to get involved in solid waste disposal and waste-to-energy issues. Private business members were included from the start, he said, because it was clear that private enterprise would be playing a major role in garbage disposal. The conference itself does not provide funds to the group, he said; its entire budget, which he would not disclose, comes from member dues.

The group operates both as the National Resource Recovery Association and as part of a coalition on resource recovery and the environment that also includes other groups from the private and public sectors. Besides testifying before Congress, state legislatures, and other public bodies, the group runs workshops for public officials.

A review of some of the testimony and other materials the group has provided to Congress and other public bodies shows that it identifies itself as representing the mayors' conference and does not explicitly point out that some of its funds and half its members come from business.

Musselwhite said, however, that the group's positions are determined by its board of directors, which he said is composed only of members from the public sector. And, he said, those who deal with the group most often are aware of its diverse membership.

"We certainly never hide that fact, and I think that in any discussions we have with any congressman or any congressman's staff, it is made abundantly clear who is involved and where the position comes from," Musselwhite said.

Senator Max Baucus (D–Mont.),
who chairs the Senate subcommittee that rewrote the federal law dealing with waste disposal—and who has been lobbied by the group concerning proposed changes to the Clean Air Act that apply to garbage incinerators—said he was unaware of industry involvement in the association.

However, Baucus said, "this type of thing" happens often in Washington, and it does not affect his own consideration of various arguments put before him.

Suffolk legislator Wayne Prospect (D–Dix Hills), who chairs the energy and environment committee, said he presumed the group was industry-backed, because it uses the term "resource recovery" in its title. That, he asserted, is an industry "euphemism" for incineration. But Prospect said he is nonetheless troubled that the group highlights its affiliation with the Conference of Mayors when it lobbies and testifies. "They're just using their association with the Conference of Mayors as a mask to do some private lobbying," Prospect said.

His colleague, legislator Steven Englebright (D–Setauket), feels even more strongly. "It upsets me," Englebright said when he learned that the group—whose testimony before the legislature concerned a proposal to ban the burial of toxic ash in Suffolk—was connected to the incinerator industry.

Englebright said he was particularly concerned because the legislature gives weight to groups from outside the region, believing that such experts have no particular political affiliation and can provide unbiased views. "When we have people who are essentially Indians in buffalo skins stalking the herd, we don't even know that we're being stalked," he said.

and twisted skeletons of charred metal. Dark flecks of burned garbage fly through the air in the burning chamber above the fire and ash.

Those airborne particles are called fly ash and account for about ten percent of the ash produced in an incinerator. The rest—called bottom ash—is composed of the larger chunks and unburned material that collect at the bottom of an incinerator's burning chamber.

Fly ash consists of smaller particles than bottom ash and has proportionally more surface area to which pollutants can adhere. Thus it contains higher concentrations of toxins and is more likely to exceed hazardous waste limits. Some environmentalists believe that fly ash and bottom ash should be disposed of separately, with fly ash going to a hazardous waste landfill if it contains unsafe levels of pollutants.

"Most incinerator operators indiscriminately mix fly and bottom ash," said Allen Hershkowitz, a New York–based environmentalist who has studied incineration practices around the world. "That contaminates all the ash."

Incinerator ash usually is mixed with household garbage in dumps. Environmentalists fear that this multiplies the pollution threat because chemicals in the ash could mix with garbage to speed the leakage of toxins into groundwater. Ash dumped in separate landfills also eventually leaks through liners into the ground, and communities are extremely reluctant to take that risk.

For example, in Vermont in 1987, a state judge halted digging at a planned ash dump on the edge of a trout stream because it had no local permit. With nowhere to put ash from a brand-new incinerator in Rutland, the plant stopped burning garbage, said Mayor

Jeffrey Wennberg. Rutland's incinerator remains closed and its operator, Vicon Recovery Systems Inc., has gone bankrupt. Vermont officials have granted Rutland extended use of its landfill.

In 1985, Long Island towns began to fight to avoid the responsibility of finding a home for ash. In December of that year, when a state advisory commission targeted a 230-acre site in Yaphank, Brookhaven Supervisor Henrietta Acampora vowed that her town would "never become the garbage capital of Long Island."

The town put aside $1 million to fight the state in court. The issue was settled in 1986, when the state agencies proposing the ashfill could not get the support of Governor Mario Cuomo, and the State Legislature refused to approve the funds to buy the site.

In the interim, other towns are proposing ways to dispose of their own ash temporarily. North Hempstead wants to open a dump next to the proposed Port Washington incinerator. Hempstead will send its ash to landfills in Buffalo, Pennsylvania, and Ohio. New York State is weighing Babylon's plans for an ash dump in Wyandanch. And in November of 1987, the Appellate Division of State Supreme Court upheld a ruling that would allow Islip to dump ashes temporarily in a designated section of the town's Blydenburgh landfill in Hauppauge.

State and federal tests show that more than half the ash samples tested nationwide contained unsafe levels of lead and cadmium. But New York State environmental officials support the EPA's move to exempt ash from classification as hazardous waste. The state argues that the ash can be buried safely in landfills.

At stake is whether ash is a serious enough threat to warrant the high cost of disposing of it as a hazardous waste. "The EPA's intent regarding ash as special waste is consistent with our position," said Thomas Jorling, commissioner of the state Department of Environmental Conservation. "If you treat it as hazardous waste, that drives up cost—we think unnecessarily."

If ash were considered hazardous waste, it would have to be buried in a landfill with specially reinforced liners. New York State has only two facilities licensed to take hazardous waste, and they are unable to handle the demand.

Some scientists have called on states to require regular ash tests and to treat ash as hazardous when it fails them. Rep. James Florio (D–N.J.) proposed a bill to require the EPA to set nationwide standards for testing and disposing of ash.

Tests at one New York ash dump—the Sprout Brook landfill in Westchester County, which receives 165,000 tons of ash a year from the Peekskill incinerator—showed that liquid seeping from the landfill was more contaminated than predicted by a consultant to the plant operator.

Generally, pollution occurs as rainwater seeps through the ash and forms a liquid called leachate. Tests of the Sprout Brook leachate conducted by a private consultant in 1986 and 1987 showed that pollution levels were higher than a pilot test had projected. The number of solid particles—to which dioxins and other pollutants adhere—in the leachate was more than five times higher. And lead levels were almost twice as high. Despite the higher levels, the leachate did not exceed the EPA's threshold for hazardous waste.

Ash on the Horizon

Ash expected from LI and NYC incinerators when all plants that currently have a site are built and operating at capacity

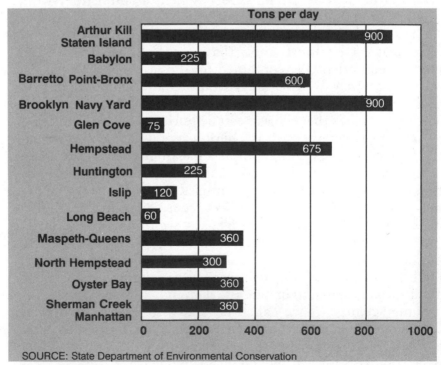

SOURCE: State Department of Environmental Conservation

Newsday/Steve Madden

The EPA said it would release its ash disposal guidelines in January of 1988. But the agency has dropped that plan preferring to wait on Congress. "We sort of backed away from that because the law governing how ash should be managed was somewhat ambiguous. We hope Congress will clarify that," said Rita Calvan, spokeswoman for Porter, EPA's top garbage official.

While the EPA has withdrawn from issuing guidance to local officials, the potentially toxic ash that became part of Long Island roads remains untested.

"It's unbelievable. There's every reason to think that very severe environmental problems could occur," said Denison of the Environmental Defense Fund. "Nobody can know that the ash is safe without a great deal of testing."

State tests at six incinerators in 1987 showed that more than half the samples—including ash from the Glen Cove incinerator—exceeded federal thresholds for hazardous waste.

Department of Environmental Conservation officials said they knew that ash had been sprinkled across Long Island over the course of a year and a half but had no authority to stop the practice.

"When you use ash as a raw material for something else, it's no longer a waste, so we don't regulate it," said Gerald Brezner, regional solid waste engineer for the state DEC. "From what I know of it, I'm fairly comfortable it doesn't leach."

Federal officials expressed more concern. "You seem to have a problem on Long Island," said David Tanner of the EPA inspector general's office. "You really would need to test the ash that was used to see what it actually is leaching before you would know whether it's safe."

Contractor Schleider, who used his own recipe to mix the ash with concrete for road base, does not know where most of it went.

But he does remember selling several tons directly to the Town of Huntington, which spread the ash-laced mixture under asphalt along a one-mile stretch of Town Line Road that runs past a dozen single-family homes and the town landfill.

Huntington Highway Superintendent Henry J. Murer said he did not know that the road base—which cost $12 a yard—contained ash. "We may have purchased crushed concrete from him for the job, but I don't know about whether it had ash. I can't confirm that," Murer said.

WHEN NOT SEEING IS BELIEVING
By Michelle Slatalla

The two state inspectors had come to look for smoke. They stood a few yards from the Baltimore garbage incinerator, glancing from their watches to the smokestack every 15 seconds. The sky seemed smoke-free, the only visible emissions coming from cars whizzing by on nearby I-95. So, after 15 minutes, they drove back downtown and wrote a report that recommended a one-year renewal of the plant's license to burn.

The report on the incinerator—which is widely considered one of the safest in the country—did not mention emission rates for such common pollutants as lead, acid gases, or dioxin. That is because the State of Maryland, which has allowed the plant to operate for more than two years since such tests were last conducted, does not require regular tests. In fact, the plant's dioxin emissions have never been measured.

"How do we know the plant is still meeting its permit requirements? That's a good question," said Justin Hsu, one of the inspectors for the Maryland Department of the Environment. "The only way to know for sure is to go back and test again or to install a monitor to get daily or hourly emissions data.

"We can see that the stack is so clean that we can't see anything coming out," Hsu said. "That's why we're satisfied." But because most air pollutants are invisible, some states—including Oregon and Massachusetts—require annual testing for specific pollutants. A bill pending in Congress would force the federal government to set limits and test regularly for two dozen pollutants.

Town Line Road resident Carl Foglia's front yard was dappled by drips of asphalt left behind when the town repaved the street. "I've had no problems, but maybe they should test the ash," Foglia said.

Down the block and across the street, neighbor Gordon Newman said, "If you tell people around here that some ash was used in the road, they don't know what that means."

Glen Cove officials say the plant operator has a contract to send the ash to a Buffalo landfill. But Steve Passage, president of Montenay Power Corp., the plant operator, said the ash is going to Resource Processing Corp. of Delaware, to be processed into roadbase. Resource Processing officials refused to say where the ash is being used as roadbase now.

Uncertainty about air pollution does not stop owner Signal-RESCO from boasting that the incinerator is a state-of-the-art model of environmental safety and efficient burning.

"We don't speak of this plant as an incinerator, although incineration is what we do. You wouldn't call Maxim's a lunchroom," said plant tour guide Lou Demely. "And you wouldn't call the Waldorf-Astoria a rooming house, would you?"

Indeed, with a thin, vertical steel sculpture on its front lawn and a three-story fisheye window that winks at cars passing on I-95, the plant looks as if it works well.

Built on the site of an older incinerator that never worked properly, the mass-burn plant sits on the edge of Westport, a working-class neighborhood whose row houses are home to about 5,000 people. While residents such as Demely—who works part-time at the plant—welcome the incinerator as a neighbor, others say they wish the state kept a closer eye on the operation.

"We don't complain much around here, but it [the smell] takes your breath away when the full trucks go by," said Mary Cartright, who has lived on Annapolis Road a few blocks from the plant for ten years. She blames the plant for "a lot of dirt on our stoops and on our clothes. We wonder what's really coming out."

She looked toward the plant. And like the state inspectors, she saw no smoke coming from the stack.

Researchers at the State University at Stony Brook have been trying, since 1985, to discover whether ash can be used safely as a building material. They are perfecting a blend of ash, concrete, and water that will bind the pollutants in ash so they will not leach. The researchers built an underwater fishing reef of ash in Long Island Sound and are constructing an ash-block boathouse on campus. "We think we have a recipe that binds the ash," said Frank Roethel, a professor at the Marine Sciences Research Center at Stony Brook. "But we're still testing."

The Town Line Road residents said they would like to see the results.

OREGON FITS EMISSION LEVEL TO THE PLANT

By Michelle Slatalla

To its builders, the incinerator on the edge of farmland in Brooks, Oregon, is a nationwide model of environmental safety. But when tests conducted before the plant opened showed that it pumped more pollution into the air than the state allowed, officials decided to bend the rules.

In 1986, tests at the brand-new Ogden Martin Systems Inc. incinerator, in Marion County 35 miles south of Portland, showed that, while no smoke was visible from the stack, the plant exceeded its emissions limits for nitrogen oxides, odorless gases that contribute to smog and acid rain and, if inhaled in heavy doses, irritate the pulmonary system. Despite the findings, the state did not require changes in the design of the 550-ton-per-day mass-burn plant. Instead, the state raised the emissions limit—by 30 percent.

State officials say the nitrogen oxide emissions are greater than expected because the state requires the incinerator to burn garbage at a high temperature to try to avoid creating dioxins, which cause cancer in rodents and may cause cancer in humans. At hotter temperatures, more nitrogen oxides are formed.

"It is a trade-off," said Wendy Sims, a senior engineer with Oregon's Department of Environmental Quality. She said the state allowed the increase in emissions of nitrogen oxides—from 94 pounds an hour to 122.2 pounds an hour—only after officials determined the increase in pollution would not be harmful. "The difference with the increase was very, very small," Sims said.

But environmentalists say the state should have done more comprehensive studies to calculate the increased risk to human health. "They set standards in the first place to protect health, then they change the standards without telling us what the change means," said Daniel Wartenberg, a professor of environmental medicine at Rutgers University in New Brunswick, New Jersey. "Do they think people are stupid, that they just won't notice?"

And David Schreiner, who operates a flower farm billed as "America's Iris Headquarters" about a mile away from the plant, worries that the acid gases emitted from what he calls "that burner" may prompt the development of plant-killing fungi that attack irises. So far, however, Schreiner said he has not seen any effect on his flowers. "Incineration is a quick-fix solu-

David Schreiner and his fields of irises, about a mile from the garbage-burning plant in Brooks, Oregon.

tion to a political problem that municipalities have with garbage," Schreiner said. "But I have to live with the quick fix. And I worry that in the long run it won't be a solution after all."

In summer, 1987, tests showed that the plant was emitting too much sulfur dioxide, which contributes to acid rain. But the state allowed the incinerator to continue burning while the cause of the increase was investigated. Sims said the plant's operators traced the excess sulfur dioxide emissions to large loads of wallboard. As a result, plant operators will "screen the loads of garbage that come in and ban that material from the plant," Sims said.

Plant manager Fred Engelhardt said that there is no need for concern and that tests show the plant's emissions rates are among the lowest in the nation. A federal Environmental Protection Agency study released in 1987 showed that the plant's emissions rates are below most other plants for which data arc available.

"When someone says 'incinerator,' what comes to mind is something belching black smoke and trucks coming in and out in a frenzy of noise," Engelhardt said. "That's not the reality. This plant is a model for the rest of the country."

Engelhardt said he has encouraged Brooks residents to tour the plant to see how clean it is. Concerned with the image of the incinerator—and of the garbage-burning industry in general—Ogden Martin officials also have tried to soothe the fears of nearby residents in other ways.

Flower beds filled with thousands of purple and yellow spiky-leaved irises surround the plant's parking lot. Ogden Martin officials bought the flowers at America's Iris Headquarters.

PART V

THE
NEW YORK
EXPERIENCE

CHAPTER 9

LOCALS LEFT HOLDING THE BAG

BY MARIE COCCO

More than a decade ago, a cadre of lawmakers and environmentalists joined forces to try to head off what they envisioned even then as a national nightmare: mountains of garbage threatening the environment.

Armed with studies predicting a dangerous shortage of landfill space and tantalized by technology that could turn waste into energy, they pushed through federal and state laws aimed at developing a rational plan to prevent a crisis.

Yet, as the nation's garbage-disposal problem grew to mammoth proportions in the 1970s and 1980s, federal and state regulators all but abdicated the roles they had set for themselves. They stood back, leaving municipal officials alone in their efforts to deal with the environmental and technological challenges—as well as the tough political decisions required by the crisis.

At the same time, the federal government indirectly subsidized garbage incinerators—a disposal method that critics say discourages recycling and could harm the environment. It granted gen-

NOTE: Jane Fritsch contributed to this story.

127

erous tax benefits to financial backers and operators of waste-to-energy plants and required utilities to buy power from them.

In New York State, officials also accelerated the push to build incinerators. In 1983, the legislature passed a law that forced the shutdown of most Long Island landfills by 1990—a deadline some officials now say may not have been justified by the short-term threat actually posed by dumps to the public drinking water supply. And despite legislation requiring state environmental officials to produce a long-term plan to handle the garbage problem, it took seven years—and legal prodding by environmentalists—for the Department of Environmental Conservation to develop the plan.

On the federal level, programs by both the Environmental Protection Agency and the Commerce Department that were to help ease the garbage burden for state and local governments, as well as push business into doing its share of recycling, were never fully funded or staffed, according to budget records and interviews with current and former federal officials. And most were dismantled after President Ronald Reagan took office in 1981.

Even though there are nearly 100 incinerators now in operation, the EPA still has not laid out definitive standards on how ash, which often has pollution levels that would make it classifiable as a hazardous waste under the agency's own tests, should be handled.

Regulations meant to force federal agencies into buying recycled products—a mandate enacted in 1976—are just now starting to take effect and are under attack from environmentalists as too weak.

"It really is a history of almost incompetence," said William Kovacs, a Washington environmental lawyer who was chief counsel to the House Subcommittee on Transportation and Commerce when the landmark 1976 law on garbage was written.

Senator Max Baucus (D–Mont.), who chairs the Senate subcommittee that helped to rewrite the law governing solid waste disposal, assessed the performance of federal officials this way: "They have not lived up to their responsibilities."

Gordon Boyd, former director of the New York State Legislative Commission on Solid Waste Management, said it was only recently that Albany's legislative leaders "looked up and recognized that nobody had been paying attention to the solid waste problem.

"We had regulations," Boyd said, "but there were no cops on the beat."

Without guidance from the federal and state governments on such issues as recycling, eliminating environmentally harmful packaging materials, and the potential hazards of incineration, local officials have been left to make decisions—and take risks— largely on their own.

Islip Supervisor Frank Jones, whose town has been forced to close one incinerator, switch to landfilling, then stop landfilling and build another incinerator, chafes at the lack of clear direction from higher levels of government. "We were all left dangling," Jones said.

Both federal and state officials defend their performance. While laws passed in the 1970s were aimed at putting in place a logical and environmentally sound garbage-disposal system for the future, by the end of the decade the more urgent—and politically explosive—issue of cleaning up toxic waste dumps took precedence, they said.

EPA officials shifted virtually all the agency's funds and personnel out of garbage-disposal programs and into hazardous-waste enforcement efforts.

"The country's focus turned to the Love Canals and valley-of-

Gordon Boyd, former director of the New York State Legislative Commission on Solid Waste Management, is now a garbage industry consultant.

THEY DUMP BY NIGHT

By Ron Davis

In 1986, a city Sanitation Department inspector surprised developer Max Berman with a citation for illegal dumping. Broken concrete was scattered across a quarter of the Staten Island wetlands site where Berman was building 100 single-family homes.

Berman says it was news to him. Only days before, the site had been clear. As it turned out, according to Berman and city officials, the concrete had actually been deposited by Colt Rental and Leasing Co., a Brooklyn demolition and trash carting firm, working on adjacent property.

"They started dumping on their site and just kept encroaching on us," said Murray Berman, son of the developer. "We didn't know about it until the Sanitation Department cited us." The issue was settled when the Bermans agreed to restore some of the wetlands and allow city engineers to conduct borings to determine if there was any garbage mixed in with the debris, according to a Sanitation Department attorney.

While Berman says the hard evidence of illegal dumping was a shock to him, it was no shock to the city. Sanitation Department officials say dozens of similar instances of illegal dumping have been repeated all over the city by Colt and other Brooklyn-based demolition and waste hauling firms controlled by brothers Eddie Sr. and Emanuel Garofalo.

Officials say the Garofalo firms are prime examples of the dozens of private dump-by-night carters who deposit hundreds of thousands of tons of garbage, old tires, and assorted trash on public streets, private construction sites, and vacant lots, daring the 62-member Sanitation Police Department to catch them in the act.

Since 1984, they say, the brothers have become the city's worst illegal dumpers. Despite hundreds of thousands of dollars in fines levied and property confiscated, officials say, the Garofalos continue to flout the law and thrive, as the waste-hauling business booms in New York City.

"He's been a persistent problem," Daniel Millstone said of Eddie Garofalo, Sr. Millstone, environmental counsel in the inspector general's office of the city Sanitation Department, added: "He's certainly way up in the top ten hit list, probably in the top three as far as egregious violations of the sanitation code."

Repeated telephone calls by

Newsday to the Garofalos' firms, which are battling the city in court, and to their attorney, Lewis Novod, were not returned.

The Garofalos are among 16 defendants in a racketeering indictment involving illegal dumping of construction debris. The New Jersey attorney general is prosecuting them and other city carters as co-conspirators in a scheme that allegedly paid more than $100,000 in bribes to the North Bergen deputy police chief and other New Jersey officials to accept city construction debris at landfills.

Despite all this, Garofalo firms still get work as subcontractors on city projects, although city officials are not able to determine to what extent.

The incentive for the sort of illegal garbage disposal attributed to the Garofalos is that it is more profitable for haulers simply to get rid of their loads in a vacant lot rather than to pay to dump it in a public or private landfill.

A 20-cubic-yard trash container can cost $375 or more in daily rental, sanitation police say. Add on the dump fee of $18.50 per cubic yard at the city's Fresh Kills landfill in Staten Island—or the much higher fees elsewhere—and waste disposal becomes a high-overhead business.

"Not everybody's a crook, but illegal dumping is sure profitable enough for everybody to be one," said Vito Mennella, a plainclothes sanitation police officer.

It's so profitable that many carters and individuals are willing to risk the fines, which range from $600 to $12,500, and impoundment of the dumper's vehicle. Mitran Equipment Corp., a Garofalo-run firm, was hit with more than $26,000 worth of fines, in addition to the impoundment of vehicles between October, 1983, and February, 1985, but still operates. The number of vehicles impounded annually for illegal dumping rose from 787 in 1983 to 1,157 in 1986, and that record was eclipsed in 1987, with 1,222 impounded.

Sanitation workers remove more than 200,000 tons of illegally dumped garbage from vacant lots annually. Other debris is mixed with dirt and crushed rock meant to serve as "clean" fill for construction foundations. "Some dumpers mix garbage with clean fill, then the garbage rots, settles, makes methane gas, and can cause cracks and other problems with your foundation," Mennella said.

Thomas O'Brien, a sanitation

police officer who, with his partner, James Tortorice, clocks 100 miles nightly patrolling for illegal dumpers, explained the techniques that make them hard to catch. "Sometimes they'll steal a U-Haul. They'll have a spotter car looking out for us. They'll use different locations because they know that if they hit the same spot two or three times in a row they'll get nailed."

Authorities say the Garofalo firms are difficult to prosecute on illegal-dumping charges because they operate under several corporate names. For example, in 1981, officials charged two Garofalo-controlled companies with dumping construction debris in a Staten Island wetlands area but dropped the charges when they could not determine which Garofalo owned the companies and trucks in question.

"They found out that if they're found guilty three times within eighteen months the vehicle involved is confiscated," said Edward Arso, captain of the sanitation police. "Every time they had a vehicle impounded twice they changed their name so they could get a brand-new start," he said.

Since the fall of 1984, said Millstone, the Garofalos' run-ins with the department have grown to "alarming proportions," with Eddie Garofalo, Sr. being cited more than three dozen times on charges of illegal dumping.

Millstone said a new computerized system of tracking contracts is expected to allow the city to identify the city contracts held by Garofalo firms in an attempt to bar them from future municipal work. "We don't take the view that he should be driven out of business," Millstone said, "but if we do take that stand against anyone, it will be someone like Eddie Garofalo."

the-drums and so on," EPA deputy administrator A. James Barnes said in an interview. And the agency's efforts on ordinary garbage, he said, "literally dried up." In New York, officials said, the Love Canal toxic-waste dump in Niagara Falls so occupied environmental officials that they had no inclination to worry about ordinary garbage.

But even after federal officials had both a strategy for handling toxic wastes and the Superfund program to help pay for cleaning up hazardous dumps, they did not turn their attention back to garbage. Environmentalists and congressional aides said that, while the program seemed to suffer from benign neglect during the Carter administration, the Reagan administration—with its disdain for federal intervention in local issues and its cutbacks in domestic spending in general—launched an all-out assault on the EPA's garbage-related activities.

Federal cutbacks had a dramatic impact on New York State's garbage-disposal efforts, Albany officials said. In 1979, the state Department of Environmental Conservation had 96 staff members assigned to nonhazardous waste programs. Four years later, that figure had plunged to 17 as a result of the federal cuts. An Assembly committee report called the cuts a "major setback" at a time "when the state needed municipal waste funding the most."

A review of federal and state involvement in garbage found that:

- Federal officials slashed the EPA's budget for garbage and other "nonhazardous waste" programs. Between 1979 and 1981, the budget was reduced from $29 million to $16 million, and the staff was cut from 128 to 74 employees. By 1982, with Reagan administration budget cuts, the staff for garbage-related projects had been cut from 74 to one, and funding had dwindled to $322,000. The EPA in 1987 had ten employees working on garbage programs, and the budget was up to $2.3 million.
- An EPA program that gave grants to states to develop long-range plans for garbage disposal was drastically scaled back in 1981 and eliminated the following year. Without the carrot of federal assistance, New York and other states ceased work on plans. "When the funding was cut, we saw a significant decline in state programs," said John H. Skinner, an EPA official who was assigned to solid waste programs in the 1970s but who now handles other issues. By 1987, only about half the states had garbage-disposal plans approved by the federal government.
- The EPA's efforts to provide technical assistance to local officials were all but eliminated. There once were 80 employees on the agency's public awareness staff, and another 100 scattered throughout the agency who performed similar functions, according to Thomas F. Williams, a retired EPA official who ran the program. Under Reagan, the public awareness staff was cut to 12. "Some of the elementary methods that were used over the years to help keep municipal governments in the loop were done away with," Williams said.
- Neither the EPA nor the Commerce Department, which were supposed to push the vast federal bureaucracy toward buying recycled products and prod industry to do the same, made a sustained effort to do so. Although these requirements date to

1976, the EPA published its first guideline for recycled products in 1983—a regulation, pushed by utilities, calling for the use of coal ash in cement bought by the government. Initial guidelines for the purchase of recycled paper were issued in October, 1987, and so far call only for federal agencies to buy recycled goods if there is a tie bid. Another proposal that would promote the use of recycled rubber in road-paving materials was quietly pulled back after states—the largest purchasers of road materials using federal funds—complained about the paperwork requirements, EPA and environmental sources said.

- The Commerce Department, which was supposed to promote resource recovery as a private enterprise and to push businesses to develop ways of reusing and recycling materials, never earmarked funds or staff to undertake such an effort. In the late 1970s, a small group of department officials researched waste-to-energy methods. But this limited work was abandoned in 1982, and the department as of 1987 had only one individual working part time on resource recovery and waste issues. "There's no money involved in this," said J. B. Cox, the Commerce Department official in charge of resource recovery. "What we're talking about is a program of a half person—half my time."

- The state DEC, which has broad authority to regulate garbage disposal through its power to issue permits for incinerators and landfills, as well as regulate their environmental performance, failed until March of 1987 to come up with the long-term garbage disposal plan first called for in 1980. Officials blamed the delay on federal cuts. But, in fact, they eventually produced the plan without money from Washington after legal prodding from environmentalists. The blueprint expresses clearly how the state feels about getting involved in garbage: "Solid waste management is local government's responsibility," it says.

- Despite the state's policy of staying out of local garbage-disposal issues, it was the legislature that made perhaps the most crucial garbage-disposal decision of the past decade. With its 1983 law ordering the closure of Long Island landfills by 1990, the area's rush to incineration began in earnest. The law was enacted because of fears that landfills were polluting vital underground water supplies. But, some state environmental officials and others now say, there is little evidence that public drinking water supplies have yet been contaminated by pollution from landfills—although experts predict such contamination will eventually spread to the public wells.

Congress never intended for the federal government to get directly involved in garbage disposal. In the Resource Conservation and Recovery Act of 1976, Congress set out to establish a nationally coordinated strategy to ensure that landfills were safe or phased out altogether, waste-to-energy technology was used appropriately, and recycling was begun on a large scale.

The EPA was to take the lead in helping municipal and state governments implement environmentally sound waste-disposal practices. The Commerce Department was to assess the merits of the emerging waste-to-energy technology and help transfer it to the public sector. And both agencies were to promote recycling.

But while federal officials defined their own roles in regulating garbage disposal narrowly, they have not lived up even to the limited mandate they carved out for themselves.

For example, panels of experts—including technicians, lawyers, waste planners, and others—were to be set up to advise communities on resource recovery methods. Like many of the other initiatives called for in the law, these panels never functioned as intended. According to EPA budget records, the panels existed for three years and issued their last report to Congress in 1979.

Senator Baucus of Montana, who said he plans extensive re-writing of the law's garbage-related portions when Congress takes it up in 1989, said that if the federal government had lived up to its mandate, its efforts might have eased the frustrations of local public officials—and private citizens—as ordinary garbage was transformed from a local nuisance to a national headache.

"It would have defused some of the present anxiety and also some of the frustration among a lot of folks around the country," Baucus said. "It wouldn't have solved everything, but it would have helped."

While it avoided being cast in the role of a regulator, the federal government was instrumental in one aspect of the garbage problem. For a variety of reasons, including the energy crisis, it created tax breaks and other indirect subsidies that the incinerator industry was able to capitalize on.

Beginning in the late 1970s and continuing through 1986, waste-to-energy investors could take advantage of energy tax credits, the investment tax credit, and depreciation rules that allowed the plants and equipment to be written off much faster than they actually wore out. Even if the plants were privately owned, special provisions of the tax law allowed tax-exempt bonds to be used to finance them.

The tax changes enacted by Congress in 1986 tightened many of these rules, but incinerators still enjoy some tax advantages.

And yet another large subsidy remains: Under a 1978 law meant to encourage the development of alternative fuels, utility companies must buy the power generated from garbage-burning plants. The price they pay is set by states—a price that in some cases exceeds what it would cost the private utility to generate that power on its own.

New York State officials, meanwhile, blame federal cutbacks for much of their inaction on solid waste. But legislative watchdogs and environmentalists say the state, too, is at fault.

A 1980 state law gave the DEC far-reaching responsibility to promote resource recovery in New York and to provide state aid and technical help to localities for resource recovery projects and other solid waste management programs. But five years later, an Assembly Ways and Means Committee report concluded that, except for providing some funds to New York City, the DEC "has not implemented this law."

Inadequate funding for the entire department, the report said, has hobbled it. Between 1970 and 1982, the DEC's budget increased from $33.1 million to $59.2 million—minimal growth that, when inflation is factored in, actually amounted to a cut of more than 27 percent. The number of enforcement officers was cut, even as the department's responsibilities were burgeoning. Even deliberate efforts the state made to grapple with the garbage-disposal crisis have been beset with difficulties.

The 1982 bottle-deposit bill, envisioned as both an anti-litter initiative and a linchpin in the state's bid to reduce the volume of waste thrown into landfills, has been poorly enforced, environmental critics charge. A 1985 report by the New York Public Interest Research Group found that, statewide, almost half the retailers who are supposed to accept returnable containers have imposed at least one illegal restriction on consumers seeking to return cans or bottles. And, the group contends, the DEC has done little to force merchant compliance.

Department officials acknowledge that they added no enforcement staff to monitor adherence to the law and only temporarily
shifted employees to that task at the outset of the bottle initiative.

Thomas Jorling, who took over as DEC commissioner in 1986,
said the department has not done all it should to resolve the
garbage predicament. "I think the state has been inadequate and
the local governments have been inadequate," he said.

In the wake of Long Island's wandering garbage barge, floating
hospital waste on the New Jersey shore and other incidents that
dramatized the magnitude of the garbage mess—and increased
the political risk of inaction—officials in both Washington and
Albany say they intend to redouble their efforts. But they offer
few concrete proposals and insist that local governments—despite their relative lack of financial resources and expertise—
must continue to bear most, if not all, responsibility.

DEC spokesman R. W. Groneman, calling 1987 "the year of
garbage," pointed to the creation in September of that year of a
separate solid waste division as evidence of the department's
concern. The number of staff members assigned to work on nonhazardous waste, which had fallen to 17 in fiscal 1982–83,
reached 97 in 1987, even without restoration of federal grants,
Groneman said. But so far, the only initiatives the department
plans to undertake are a study of how to rid the waste stream of
potentially hazardous residues from garbage such as used batteries and newspaper ink, and a renewed push for legislation to
force bottlers to turn over $60 million a year in unclaimed bottle
deposits to the state. The state, in turn, would transfer the money
to local governments for use on waste-disposal programs. The
bottle-deposit measure failed in 1986 amid heavy lobbying by the
beverage industry, officials said.

Any larger role the state undertakes will be governed by the
solid waste management plan—finalized only in 1987. The plan
calls for reducing the waste stream through recycling and other
methods, construction of waste-to-energy plants, and landfilling.

"The vision that I have is that we will become a much more
conserving and recycling society," Jorling said. "It's going to happen because the costs of disposing [of garbage] are going to go up
dramatically."

But NYPIRG official Walter Hang contends that the state's plan "suffers from a paradox of purpose." While it says that waste reduction and recycling are top priorities, Hang said, the document is bereft of detailed proposals needed to establish recycling programs. At the same time, garbage incinerators—which do nothing to reduce the amount of trash thrown away—are indirectly subsidized by the state because they are, by tradition, financed with tax-free bonds.

State law gives New York officials the power to help shape crucial decisions about the construction and operation of garbage-burning plants. In New York, incinerators must have two state permits to operate. Before those permits can be issued, the state must approve the plant's site and design, and the operator's plans for groundwater monitoring, ash disposal, and air pollution control. If the state is not satisfied with the conclusions drawn by cities and towns, it can ask local officials to redo the work—an option the state has never invoked.

In Washington, the EPA has begun to stir.

The agency's 1989 budget devoted to garbage includes $13 million for contracts, grants, and research as well as a staff of 64 people. Yet at this point, officials said, the EPA is still sorting out what the federal role should be.

"It's clear that there is a very major environmental management issue there," Barnes, the EPA deputy administrator, said.

Porter said he would like to require, rather than suggest, that states submit solid waste management plans. But neither official envisions the reinstitution of federal grants to help the states.

In October, 1987, Barnes called together recyclers representing both government and industry for what participants in the meeting said was the first time in memory that the agency displayed interest in their activities. Barnes himself said he was surprised to learn how extensive recycling has become in some states, localities, and industries, and acknowledged that "a lot of that effort had really passed the federal government by."

Meanwhile, EPA officials are still working on new regulations for municipal landfills that were first ordered by Congress in 1984. The agency expects to issue new landfill rules in the fall of 1989, after which states will have 18 months to adopt them.

The EPA also is about to issue guidelines—but not enforceable regulations—for the disposal of ash from garbage incinerators. The ash often has levels of contaminants that make it classifiable as a hazardous waste under the EPA's own tests. But Porter, to the chagrin of environmentalists, has said the agency will not require the residue to be disposed of as if it were hazardous, and instead will suggest that it be treated as a "special" waste with its own, less stringent, disposal requirements. The agency argues that Congress intended specifically to exempt the ash from municipal trash incinerators from strict hazardous waste disposal rules.

There are bills pending in Congress that would force stricter ash handling procedures. And both Baucus and aides to Rep. Thomas Luken (D–Ohio), who chairs the House subcommittee helping to rewrite the federal law governing garbage disposal, said the ash issue is likely to be addressed then.

Years of EPA inaction—and pressure on members of Congress from home-state officials—may lead lawmakers to write into the law their own minimum environmental standards for landfills and garbage incinerators. It is also possible, Baucus said, that some kind of mandatory recycling provisions could be enacted.

The EPA, however, is unlikely to support the notion of Congress writing its own environmental standards into legislation. And as for recycling, Porter said, he is unsure whether it is useful to put such mandates into federal law.

As the agency regroups for some still-unspecified role in managing the nation's garbage crisis, Barnes said he does not foresee any direct federal involvement. Rather, he said, it is more likely that the EPA will increase its research and provide technical help to localities. Those are precisely the jobs Congress first assigned it more than a decade ago.

CHAPTER 10

WILL IT GO UP IN SMOKE?

By Ron Davis and Barry Meier

While country kids wondered what made the sky blue, Vito Mennella's childhood imagination grappled with a somewhat different cosmic question. "When I was growing up in Brooklyn, I always used to wonder why vacant lots were higher than the sidewalk," Mennella recalls.

These days, the vacant lots are getting still higher. And Mennella, an undercover Sanitation Department detective, now knows the reason: illegally dumped garbage. "Some guy who doesn't want to pay a landfill fee will just pull up to a lot and let loose," he says.

New York City, the nation's largest municipal garbage producer, is starting to choke on trash. As nearby landfills in New Jersey and Long Island close or sharply boost dumping fees, millions of tons of city-spawned trash, garbage that once would have been buried elsewhere, is staying in the city.

Huge tractor-trailers jam-packed with trash are being abandoned along desolate city streets. Carters are secreting rotting garbage in the foundations of homes and commercial buildings.

And the rate of refuse legally dumped into the already-clogged heart of the city's sanitation system—the giant Fresh Kills land-fill on Staten Island—was up 20 percent between 1985 and 1987, a record-setting jump.

As the city presses ahead to erect the first of five major garbage-burning plants at a total cost of at least $2 billion, the battle continues over whether New York's rubbish ills are best cured by burning trash, recycling it, or, as some would prefer, simply ship-ping it out of sight.

"Until we learn to eat garbage, we're going to have to dispose of it," says Mayor Edward I. Koch.

The resolution of how best to do that, many agree, is likely to affect New York's future in ways few of the city's 7.3 million

Mayor Koch and Councilman Sheldon Leffler launch a newspaper recycling program in May 1987.

Newsday/Bruce Gilbert

residents yet imagine. The anticipated future costs of disposing of the city's trash, Koch says, may one day compete with public spending on schools, recreation, and other social services.

"There is a limited pot of money, and people will have to understand that you have to engage in triage," he says.

But even as the debate over the city's garbage crunch intensifies, New Yorkers continue to produce garbage at the staggering annual rate of more than 2,000 pounds for every man, woman, and child living here.

As the city's garbage problems mount, a *Newsday* survey of city incineration plants, their financing and sanitation policies, has found that:

- The closing of landfills outside New York is sharply cutting into the projected lifespan of Staten Island's Fresh Kills. In 1985, city officials projected that Fresh Kills, which gets 95 percent of the city's landfilled trash, might last another 14 years. Today's most optimistic estimate for Fresh Kills' lifespan: a decade.
- While the Sanitation Department publicly advocates recycling, at least one internal department document shows that city policies are, in fact, undermining recycling efforts here. A 1987 internal memo states that low dumping charges at Fresh Kills encourage area businesses "not to recycle," at the taxpayer's expense.
- City plans to erect its first new garbage incinerator at the Brooklyn Navy Yard have been bogged down since 1981 by political maneuvering, community opposition, and corporate mergers. Even when completed, the plant, the first of five to be built in the boroughs, will process only 11 percent of the city's total trash.
- Despite calls by the city's own consultant for competitive selection of bond underwriters for the Brooklyn Navy Yard incinerator project, Merrill Lynch & Co. and Goldman Sachs & Co. gained the lucrative underwriting roles without facing competition. City lawyers found that Comptroller Harrison J. Goldin's office "resisted" competitive selection of project bond underwriters, a charge a Goldin spokesman denies.
- Serious environmental concerns still loom about the hazards posed by the planned battery of giant garbage-burning plants in the city. One major worry: How to dispose of an estimated 3,600 tons of toxic ash that city officials say will be produced by the five new incinerators daily.

To be sure, there are not any easy answers to New York's garbage problem. In recent years, for example, the city has been forced to close its dumps in the Bronx and Brooklyn because of environmental problems and other woes. Also, while some Long Island communities such as Hempstead and Oyster Bay are exporting their trash to out-of-state landfills, many experts believe that many distant dumps would be loath to accept trash from New York City. "We don't have all that many options in dealing with our trash problem," says Norman Steisel, a former sanitation commissioner who is an executive with Lazard Freres & Co., a Wall Street investment firm active in promoting incinerator projects.

About 60 percent of New York's total garbage is residential, with the remainder coming from businesses, restaurants, construction sites, and other commercial operations. As elsewhere, one of the big reasons for the city's garbage boom is explosive

The Brooklyn Navy Yard, site of the first of five planned garbage incinerators.

Newsday/Audrey C. Tiernan

growth in product packaging, which sanitation officials estimate now comprises up to 35 percent of total city rubbish.

In response to the packaging problem, Koch said in December of 1987 that he would ask McDonald's Corp. and other major fast-food vendors to stop selling their products here in Styrofoam containers.

New York's changing nature is also affecting its garbage output. The rapid growth of the financial services industry has not only led to new jobs, but to mounting heaps of computer print-outs and photocopies. While Manhattan, the city's financial center, houses about 21 percent of New York's populace, it churns out about 40 percent of its total trash, city estimates show.

"Gary has steel mills, Detroit has auto plants, Nashville has recording studios, and we have paper pushers," says Brendan Sexton, Department of Sanitation commissioner.

Since 1985, the states of Pennsylvania and New Jersey, as well as some Long Island communities, have restricted the dumping of New York wastes in their landfills. The restrictions in New Jersey alone mean that some 2,000 tons of city-generated trash once dumped daily in Garden State landfills by commercial carters are now being disposed of in New York, sanitation officials estimate.

After months of heated public wrangling, the Board of Estimate in 1985 approved plans to erect a major incinerator in each of New York's five boroughs. The decision, city officials acknowledge, combined political expediency with a need to find a quick fix to the impending landfill crisis.

When completed, sometime during the next decade, the five massive units are expected to burn about 12,000 tons of garbage each day, or slightly more than 44 percent of the estimated 27,000 tons of trash produced daily. City officials hope that the incinerator program will extend the life of Fresh Kills but are also beginning to formulate plans to export garbage out of New York.

Burning New York's garbage is not a new approach. Prior to the advent of stricter air pollution laws in the 1970s, for example, about one-third of city waste was being burned in nearly a dozen municipal facilities and an added 2,500 apartment-house incinerators.

Nonetheless, efforts to build major new incinerators here have flopped for two decades. As far back as 1967, then-Mayor John Lindsay pressed for the erection of a $110 million "super-incinerator" at the Brooklyn Navy Yard that would burn 50 percent of the city's garbage. When that futuristic behemoth failed to get off the drawing board, Abraham Beame, Lindsay's successor, also unsuccessfully urged that a $90 million trash-to-fuel facility be built at the same locale.

The Brooklyn Navy Yard incinerator, proposed by the Koch administration, is the first of the planned new incinerators. Located amidst the rambling warehouses and dry docks along the East River, the plant, slated to be built and operated by Wheelabrator Technologies Inc., would burn 3,000 tons of garbage daily. Steam, produced by the heat of burning garbage, would be sold to Con Edison and the revenues split between Wheelabrator and the city.

The sites for the four other planned incinerators are: Fresh Kills, Staten Island; Barretto Point, Bronx; Maspeth, Queens; and Wards Island, Manhattan. The city is conducting environmental impact reviews at those sites.

Most of those involved in the incineration debate believe that if the city is successful in erecting the Brooklyn plant, the others will follow. "Brooklyn is the watershed project," says Allen J. Hershkowitz, director of municipal solid waste research for INFORM, a local nonprofit research group.

Construction of the Navy Yard plant has faced an uphill battle. The project's bitterest opponents have included, among others, the Hasidic community in neighboring Williamsburg, a group that continues to fight the garbage-burning project on both environmental grounds and its potential impact on real-estate prices.

"We are determined to fight this [incinerator] to the end," says Rabbi Chaim Stauber, executive vice president of United Jewish Organizations of Williamsburg.

As the fight over the planned Brooklyn incinerator has dragged on, its price tag has skyrocketed. From an initial 1981 projected construction cost of $169.5 million, the facility's cost had risen by 1985 to some $290 million, due largely to inflation and added environmental safeguards. The projected cost of both building and financing the Navy Yard project if begun today is $445 million.

The Brooklyn plan has also led to some behind-the-scenes City Hall maneuvering over the project's financial plums. Consider the case of Merrill Lynch and Goldman Sachs, which stand to net healthy shares of more than $14 million in undcrwriters' fees generated by a bond offering that is to be used to finance the Brooklyn facility.

Although Salomon Brothers Inc., the city's financial adviser, urged a competitive selection of leading bond underwriters, a city legal report found that the comptroller's office fought the effort and, instead, pushed Merrill Lynch and Goldman for the work. "The use of [a competitive review process] was resisted by the comptroller's office," the December, 1986, report held.

Officials of Goldin's office dispute the report's findings. Among other things, they claim sanitation officials themselves dropped calls for a competitive review of bond underwriters because of the need to complete contract arrangements quickly for the Brooklyn facility. "At no time did we resist the use of a [competitive process] to select underwriters," says Steven Matthews, a Goldin spokesman. Matthews also said that the comptroller's office did not choose the underwriters.

While the city legal report found that sanitation officials failed to push strongly for competition for bond underwriters, it concluded that they and other city officials "acquiesced" to the undisputed selection of Merrill Lynch, Goldman, and other underwriters favored by Goldin's office.

Deputy Sanitation Commissioner Paul Casowitz, an architect of the city's incineration projects, estimates that the bond offering for the Brooklyn facility will generate about $14 million in underwriters' fees.

Both Merrill Lynch and Goldman have been consistent contributors to both Goldin and Koch campaigns. City officials, including Koch, attempt to downplay the handling of the underwriting plums but some vow not to repeat the incident. "I think there definitely should be competition among the underwriters for the next [incinerator] project," Commissioner Sexton says.

The issues of how many incinerators will be needed and the level of public funding are matters of contention. "You don't turn to incineration before you do everything you can to avoid it," says

INFORM's Hershkowitz. Hershkowitz, like many environmentalists, says he does not strictly oppose burning garbage but believes the city is doing little to pursue recycling and other alternatives.

Statistics appear to back him up. For example, while plans call for the Sanitation Department to recycle 15 percent of publicly collected garbage by 1991, the department in 1987 recycled less than one percent of collected trash.

Budgetary outlays for recycling are also lean. For instance, although some $12.5 million alone is being spent on exterior decoration and landscaping for the Brooklyn Navy Yard incinerator, total municipal expenditures for recycling for fiscal year 1987 amounted to a mere $10.2 million.

To be sure, there is sharp debate over how much of New York City's garbage can be recycled. Some advocates of the technique contend that up to 75 percent of the city's trash is recyclable, although no U.S. city comes even close to that figure. Other environmentalists maintain that a 40 percent figure is more realistic.

Sanitation officials contend that such estimates are pie-in-the-sky. Few New Yorkers, they believe, are ready to spend time separating newspapers, glass, metal cans, and other recyclable items from their trash. They also point out that, as yet, few industrial markets exist for recycled materials. "People who make these kind of estimates just don't know what they're talking about," says Deputy Commissioner Casowitz.

Whatever the case, the future of even limited recycling in the city appears to be jeopardized by several factors, including internal Sanitation Department policies. Take the case of the price commercial carters pay to dump trash at Fresh Kills. Such landfill fees play a central role in sanitation strategies, experts say, because the quicker dumps fill up, the sooner high-priced incinerators will be needed.

But as internal sanitation documents indicate, long-standing city policies of subsidizing low commercial dumping fees at Fresh Kills with taxpayer dollars may itself be partly to blame for the city's trash crisis.

"Past years' low rates encourage[d] excessive use of the City's [landfill] facilities and provided no economic incentive for recycling or the development of alternative facilities by private industry," states a May, 1987, internal Sanitation Department document on landfill-fee pricing.

Dumping fees have climbed in recent years. The cost of dumping a ton of garbage at Fresh Kills today is now $37, compared to $6.70 a decade ago. But despite the rise, city dumping fees still lag sharply behind neighboring areas such as New Jersey, where private carters pay up to $100 or more per ton to dump commercial trash. "Until we bring our dumping fees in line with other neighboring areas, local waste generators will essentially be subsidized not to recycle," a recent sanitation document stated.

Many professionals, including Commissioner Sexton, say they believe that the current dump rate hikes still do not provide a strong enough incentive for businesses to recycle. While Sexton favors increases, he fears that another sharp climb in those fees could jar small business owners. "We couldn't do it all at once because it would have been so disruptive," he contends.

Other experts disagree. Former commissioner Steisel, for example, maintains that a quick rise in dump fees will be readily reabsorbed by the private sector. "I think that, while the cost of disposal would increase by a factor of two, it still represents a small cost of doing business," he says.

An independent economic study commissioned by the Sanitation Department supports that contention. It points out that current dumping fees, while higher, are still based on data that underestimate the cost of replacing dumps with incinerators.

Fixated by incineration's high-tech lure, the Sanitation Department, some critics maintain, has also spawned a bureaucratic culture resistant to recycling. "It's an institutional resistance, because [incinerator] technology is extremely complicated and has required a lot of planning," says Margaret Ayers, executive director of the Robert Sterling Clark Foundation, a local philanthropic trust that has funded garbage studies.

Indeed, it was only after the Board of Estimate's 1985 approval of a city-wide incineration plan that sanitation officials set up a separate recycling office within the department. Prior to that time, two department employees handled all recycling-related matters.

"Municipal leaders in general have been slow to realize the

value of recycling," says Joan Edwards, who directs the department's current 15-person recycling staff.

Such lethargy can have serious results. For example, the exact composition of New York City's waste is unknown because the city has never bothered to analyze it thoroughly. "The most recent data I've seen on city waste composition is ten years old," says Marjorie Clarke, an environmental scientist formerly with the city Sanitation Department. "And that isn't detailed enough to tell us exactly what is in the garbage so that we can develop recycling strategies," she says.

A comprehensive waste composition study is being conducted by the Sanitation Department to give operators of the Brooklyn resource recovery plant data on the makeup of its garbage fuel, said department spokesman Vito Turso. That information will

Sanitation Commissioner Brendan Sexton displays a yo-yo that he found in a garbage bale during a recent visit to the Southwest incinerator in the Bensonhurst section of Brooklyn.

Newsday/Chris Hatch

also be used by the recycling office to develop its programs, he said.

The rapid depletion of Fresh Kills is spurring some added recycling efforts. Municipal spending on recycling in 1987, for example, represented more than a three-fold increase from spending levels in the 1986 fiscal year. The majority of those funds were spent on pilot programs to collect newspapers from selected neighborhoods and on curbsides and recycled white paper from municipal and private offices. The city will also spend $4 million to renovate and reopen an East Harlem processing center for recycled bottles and cans.

A private Bronx recycling center nicknamed "R2B2" has been enlisted in the city's effort. Inside the two-story building at 1809 Carter Avenue in the Tremont section of the Bronx, 15-foot-high mountains of aluminum and tin cans, glass containers, and plastic soda bottles await processing amid the whine and moan of compactors, sorters, and grinders.

David Muchnick, president of R2B2, which stands for Recoverable Resources/Boro Bronx 2000 Inc., said the recycling center has a $260,000 annual contract with the city to accept recyclables from the public.

But while Muchnick and other private recyclers believe there is a growing market for recycled materials, others fear that, without proper city planning, incinerators may soon be burning up materials that can be recycled fruitfully.

The state has halted temporarily the city's incineration program. In November, 1988, DEC Commissioner Thomas C. Jorling refused to grant a construction permit for the Navy Yard plant because of doubts about the city's recycling and ash disposal plans.

In response, the city is expected to adopt a tougher mandatory recycling bill—one that would commit the city to recover by 1994 about 1.2 million tons of material each year.

On ash, the city and state are at a stalemate. The city had planned to dump the 280,000 tons of ash produced by the Navy Yard plant at the Fresh Kills landfill. If the city wants to pursue its plan, Jorling urged that it first excavate 60 feet of rotting waste buried at that part of Fresh Kills and then install four protective liners before burying any ash.

City Sanitation Commissioner Sexton calls Jorling's plan to dig up garbage dangerous and difficult. He says he doesn't know what ash-disposal plan will satisfy Jorling before the state will issue a construction permit for the city's first resource recovery plant.

Some garbage experts such as INFORM's Hershkowitz maintain that air pollution and ash hazards associated with burning trash could be reduced simply by ensuring that certain products never get into incinerators. And, he argues, incinerator ash, rather than being buried, can be turned into cinderblock, which is also composed of ash. Scientists at the State University at Stony Brook are currently studying the safety of such blocks.

Commissioner Sexton says that he is giving serious consideration to pushing the mandatory recycling of batteries, tires, and other nuisance wastes.

Even if the Brooklyn plant overcomes its remaining licensing hurdles, the facility, which will take three years to construct, will not start operating until 1992, at the earliest. And while five garbage-burning plants, officials say, will make a meaningful dent in New York's present problem, the $2 billion fix will only result in about seven years of added life at the Fresh Kills dump.

As a result, another generation of New Yorkers will face a day of garbage judgment. And the same battles are likely to start all over again as the city scrambles to find a way out of the garbage maze.

"People can complain about these incinerators all they want," says Sexton. "They can argue against them, they can write to editors, but in the end the garbage is going to win."

UNITY LACKING ON TRASH FRONT

By Alvin E. Bessent

With Long Island under the gun to close its landfills by 1990, incineration has emerged as the garbage disposal method of choice in half the Island's municipalities, while some others are still casting about for solutions.

But in facing an era in which garbage trucks on Long Island will not be allowed simply to lumber up to a hole in the ground and disgorge their burden for burial, local officials say that they were left struggling to fashion workable disposal plans from imperfect options and that they had limited technical expertise and little help from state experts.

In Long Beach, where the officials decided to improve and reopen a closed incinerator, City Manager Edwin Eaton said the decision was made with little effort at regional coordination, even though six towns and the city of Glen Cove have also opted to burn garbage. "We were all doing the same thing in a vacuum. It's not government functioning at its best," Eaton said.

Officials in other towns where incinerators are planned said they would informally meet and sometimes exchange public documents related to solid waste management. But they did not systematically share information or jointly analyze options. "We always used to get together at one time or another, but I don't think there was that much sharing of information," said Rim Giedraitis, president of the Islip Resource Recovery Agency.

So towns individually launched evaluations of the content of their garbage, studied environmental concerns, evaluated technologies and vendors, and made decisions on their own.

What resulted in the six municipalities were mass-burn incinerators, but with capacities ranging from half the amount of a municipality's garbage to about twice the amount, and at costs ranging from $13 million to retrofit the existing incinerator in Long Beach, to $370 million to build a new incinerator on the site of a closed plant in Hempstead. Also involved in the individual decisions were differing estimates on the amount of garbage that could be recycled. The state Department of Environmental Conservation has required that recycling be part of the waste disposal plan for every municipality seeking an incinerator permit, but there has been no guideline on the percentage of garbage to be recycled.

After studying the content of the town's garbage in 1984, Babylon Supervisor Anthony Noto said town officials concluded that about 25 percent of Babylon's garbage could be recycled. But the rest would have to be burned. "There is more than one alternative, but we had to tie it to burning," Noto said. "There is no way to get rid of garbage without burning as part of the mix."

Islip Supervisor Frank Jones said town officials, before he took office in 1987, reached the same conclusion about the need to burn some garbage. At that time, the town was planning on recycling some garbage and building two incinerators to burn the rest.

The first plant, which is under construction, will handle 518 tons a day, a little more than half the 986 tons of trash the town generates daily. Officials say the second may still eventually be built; meanwhile, the town has decided to push recycling and composting, disposal methods that are cheaper than burning and, they hope, will make the second plant unnecessary.

In the small cities of Long Beach and Glen Cove, incinerators have been built with the capacity to burn about twice the amount of garbage those municipalities generate. The excess capacity, some of which Glen Cove uses to burn gar-

bage from a number of Oyster Bay villages, is insurance against down-time, said Mayor Vincent Suozzi.

Harold Berger, the DEC regional director, said excess capacity is desirable if you evaluate the need for incinerator capacity on an Island-wide basis. "Ideally, garbage from one town, when its plant is down, could be burned in another town's incinerator," he said.

"In Riverhead, officials initially resisted the move to incineration but now hope to build a plant jointly with other towns," said Joseph Janoski, town supervisor since 1980. The town has yet to find another town as an interested incinerator partner.

"Riverhead changed course after being deluged for years with salesmen pushing garbage-to-oil, garbage-to-fuel pellets, garbage-to-plastic, and countless other disposal methods," Janoski said. "At one point, almost every meeting I went to had a vendor," he said. "We got to the point here where we just wouldn't sit down with another vendor."

The possibility of buying into a disposal method that might subsequently fail to pass DEC muster was judged too risky. "The state offers absolutely no technical assistance. They will not take the step of officially sanctioning any process but resource recovery as the solution," Janoski said. "We can't be an experimental laboratory for the State of New York."

Berger called statements like those easy excuses. "We need to stop pointing fingers and try to resolve the problem," he said. "No government on Long Island likes to be told by Albany what they must do. We felt that if we did that, we'd get just as much objection as we do now."

Berger said the state was not concerned with subtle differences in technology, as long as the plants, when operating, were within acceptable pollution levels. "If you build a black box that does what we want, it's okay," he said.

In fact, East Hampton is pleased that the state did not impose rigid requirements. "Certainly they [state officials] have a bias toward incineration," said Councilman Pat Trunzo. "But I think it's been helpful that each town has become its own laboratory and pioneered its own methods. We've gotten a lot farther a lot faster than if it were a monolithic approach.

"In a three-month pilot program, the town successfully recycled 85 percent of 100 families' garbage and is now studying ways to expand the program into the rest of the town," Trunzo said. Composting played a major role in the town's effort. "It's the least expensive method for dealing with solid waste," he said. "It's the most environmentally sound alternative." Southold, too, hopes to rely on composting and recycling.

The seven towns moving toward incineration are following the lead of Glen Cove, where garbage has been recycled or burned since 1983 in the only functioning waste-to-energy plant on the Island. The city, which recycles about 5 percent of its garbage, has no landfill and tried trucking refuse off the Island before building the plant. "But as options dwindled, city officials saw the handwriting on the wall," Mayor Vincent Suozzi said.

"I venture to say that if we had no incinerator in Glen Cove it would cost us about $3.5 million [annually] to dispose of garbage," Suozzi said. The city spent $1.28 million in 1987 to collect and dispose of garbage. But the move to incineration has left the city with one problem—what to do with the ash. The city has had short-term deals with other towns to dump ash in their landfills and with a private company that used the ash in paving. But since a deal with Smithtown fell apart in June, 1987, the ash has been trucked to Delaware.

Although confronted by the same time crunch on garbage, officials in Smithtown and Brookhaven have not made any firm commitments.

Smithtown Supervisor Patrick Vecchio said his town is considering the possibility of an agreement to burn its garbage in plants planned in Huntington or Islip. But no agreement has been negotiated, so the town may send out a request for proposals in case building an incinerator becomes necessary.

Public officials are betwixt and between. "Build an incinerator and have it go bad and you're in trouble. Don't build a plant and your government in Albany and political opponents will say you did nothing about the garbage crisis. . . . I think that's called a Hobson's choice," Vecchio said.

And in Brookhaven, officials said they are reviewing an environmental impact study completed this year to evaluate their alternatives.

The bottom line, Long Island officials said, is that the buck on garbage stops with them. Former Southampton Supervisor Martin Lang said, "We can put a man on the moon, but how do you get rid of your garbage?"

Across Long Island: Each municipality's approach to the garbage problem

Municipality/ tons per day	Current disposal	Landfill deadline	Long-term plan/disposal costs*	Recycling situation
Nassau				
Glen Cove 75	95% incinerated, 5% recycled	None (no landfill)	Recycling and incineration, using a $24-million, 250-ton-per-day mass-burn incinerator completed in 1983. Disposal costs, including carting and debt service: $1.28 million in 1987, $1.34 million in 1988. Average homeowner pays $141 a year.	Newspapers, glass, metals separated from incinerator ash.
Hempstead 2,565	78% landfilled in Oceanside, rest hauled off LI	Household waste to be banned from Oceanside landfill next summer	Town is building a $370-million, 2,505-ton-per-day waste-to-energy incinerator scheduled to open in August, 1989. Until then, garbage will be landfilled and shipped off LI. Disposal costs: $44 million in 1987, $72 million in 1988. Average homeowner pays $179.	Pilot, mandatory program for newspapers, glass, cans and appliances to begin in January.
Long Beach 82	Most hauled off LI, some recycled	None (no landfill)	Recycling and incineration. An incinerator closed several years ago because of air pollution problems has been retrofitted for $13 million and is ready to open. Disposal costs, including carting: $1.7 million in 1987, $2.8 million in 1988. Average homeowner pays $123.	Pilot program for newspapers, metals, composting. Ordinance enforcement begins in January. Glass, household toxic waste, and cardboard from businesses may be added to program.
North Hempstead 785	93% buried in Port Washington landfill; rest recycled	1990, when state landfill law takes effect	Incineration and recycling. Town plans to spend $219 million** to build a 990-ton-per-day incinerator; no contractor chosen. Disposal costs: $7.75 million in 1987 and $7.76 million in 1988; private carting costs the average homeowner $67 per year.	Recycles glass, newspapers, metals. Officials are considering building an interim processing facility for recyclables until the incinerator, which will have an area for sorting recyclables, is completed.
Oyster Bay 907	Less that 1% recycled, rest hauled off LI	Town landfill in Old Bethpage closed in June, 1986	Incineration and recycling. Town plans to build an incinerator; no contractor chosen and no cost estimate. Disposal costs: $20 million in 1987, $40 million in 1988; average homeowner now pays $195 for carting.	Metals, glass, newspapers in mandatory pilot project that has reduced waste stream by 9% in pilot area. Officials hope to make program townwide next year.
Suffolk				
Babylon 627	96% landfilled in Wyandanch, 4% recycled	1990, when state landfill law takes effect	Incineration and recycling. Town building a $121-million, 750-ton-per-day incineration plant to begin operation in April, 1989. Disposal costs: $3.75 million in 1987, $4 million in 1988; homeowners pay about $130 for carting.	Paper recycled now; glass, metals and plastics to be added in spring. Goal is to recycle at least 15 percent of waste stream.
Brookhaven 1,440	99% landfilled in Yapank; 1% recycled	1990, when state landfill law takes effect	Town officials are reviewing an environmental impact study to evaluate alternatives. No decision on incineration has been made. Disposal costs: $4.25 million in 1987, $5.1 million in 1988. Carting costs average homeowner $250.	Voluntary recycling and composting. Town officials plan to phase in mandatory recycling starting next month. Town to spend $1.5 million in 1988 to build recycling processing center, hopes to recycle 25% of waste stream.

Suffolk

Municipality/tons per day	Current disposal	Landfill deadline	Long-term plan/disposal costs*	Recycling situation
East Hampton 73	90% buried in Montauk and Fireplace Road landfills; rest recycled	1990, when state landfill law takes effect	Recycling, including composting for such wastes as wet paper and food. Disposal costs: $575,000 in 1987, $910,000 in 1988. Average homeowner pays $135 for carting.	Voluntary recycling of paper, cans. 100 families in pilot program recycle or compost 85 percent of their garbage. Officials to study extending program townwide.
Huntington 772	79% landfilled in East Northport, 20% burned in three incinerators that don't recover energy; less than 1% recycled	In dispute. State wants landfilling stopped now. Town wants to landfill until incinerator is finished in 1990.	Incineration and recycling. Town plans to build $154-million, 750-ton-per-day incinerator. Disposal costs: $5.6 million in 1987; $6.3 million in 1988. Carting costs average homeowner $204.	Mandatory recycling program in two refuse districts for newspapers from homes and cardboard from businesses. Next year, residential glass and cans.
Islip 986	91% landfilled in Hauppauge, 3.5% recycled, 5.5% composted	1990, when state landfill law takes effect	Incineration and recycling. A $57.3-million, 518-ton-per-day incinerator is under construction. Disposal costs: $10.9 million in 1987, $10.4 million in 1988. Costs covered by carting fees that average $167 per household.	Mandatory, townwide recycling of paper, metals and glass. Mandatory cardboard recycling for businesses to begin next month; glass and cans from businesses to be added later in 1988.
Riverhead 100	Almost all landfilled; the rest recycled	1990, when state landfill law takes effect	Town considering incineration, possibly in co-operation with other East End municipalities. No firm commitments made. Disposal costs: $530,000 in 1987. Carting costs average homeowner about $135.	Voluntary recycling of newspapers, cardboard, appliances. Glass and cans to be added in 1988. Officials are studying recycling options and may make recycling mandatory.
Shelter Island 14	Almost all buried in town landfill	1990, when state landfill law takes effect	$15,000 budgeted in 1938 for study of alternative disposal. Officials hope to landfill after deadline because landfill is not located over a deepwater recharge area. No plans for incinerator. Disposal costs: $104,176 in 1987, $129,700 in 1988. Average homeowner pays $135 for carting.	About 30 tons of newspaper recycled since voluntary program began in spring. Officials hope for 15% reduction in waste stream through recycling.
Smithtown 416	Vast majority buried in town landfill in Kings Park; small amount recycled	In dispute. State says landfill is full now. Town has excavated new section of landfill and wants to use it.	Considering incineration, but no commitments made. Increase in recycling planned. Disposal costs: $1.53 million in 1987. $1.9 million in 1988. Carting costs average homeowner $120.	Voluntary recycling of newspapers, glass, metals in some areas to expand townwide by end of year. Officials hope to reduce waste stream by 20%.
Southampton 175	Most buried in North Sea landfill; some recycled	1990, when state landfill law takes effect	Town officials hope to continue landfilling after deadline because landfill is not located over a deepwater recharge area. Incineration being considered, possibly in cooperation with other East End towns. Disposal costs: $1.2 million in 1987, $1.25 million in 1988. Carting costs average homeowner $135.	Voluntary recycling of newspapers, cardboard, some metals. Leaves and yard clippings are composted. Officials hope to reduce waste stream by 15% through recycling.
Southold 80	Most landfilled in Cutchogue; rest recycled and composted	1990, when state landfill law takes effect	Plans to dispose of bulk of waste through composting and recycling. Disposal costs: $684,030 in 1987, $802,976 in 1988. Average homeowner pays $135 for carting.	Voluntary recycling of newspapers, tires, some metals. Yard waste is composted. Town hopes to reduce waste stream by 30% through these methods.

* Cost of incinerators includes finances. **Estimate of total project cost computed by Newsday based on average financing costs reported by other towns.

Newsday/Hayes Cohen

Money to Burn

Projected annual average costs per household for garbage pickup and disposal in towns with incinerators under construction

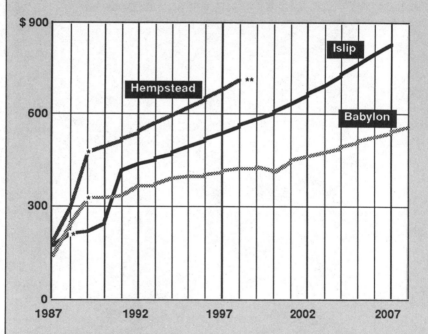

*Year incinerator is to begin operating
**Town contract to truck unburned garbage and incinerator ash runs through 1998, when contract must be renegotiated.

SOURCE: Newsday calculations based on information supplied by the towns and their consultants

CHAPTER 11

TOWN STRIKES A DEAL
IN A HURRY

By Mark McIntyre

At 6:53 a.m. on March 29, 1987, what remained of Hempstead Town's first effort at resource recovery came down in a crescendo of dynamite and plastic explosives.

In the space of six and a half seconds, as 1,000 environmentalists and thrill-seekers cheered, a 196-foot-high, steel-reinforced smokestack and two giant boilers of the closed Hempstead Resource Recovery Plant were reduced to a mass of twisted rubble along the Meadowbrook Parkway in East Garden City. A flock of birds rose up and circled frantically above the debris. Ten minutes later, as planned, a second smokestack collapsed.

The demolition was one of the few things that worked as planned for the plant. Once billed as the ultimate solution to the town's—and the nation's—garbage problem, it became a $135 million mistake, a plant that rarely worked as intended, was the subject of neighborhood complaints, and was closed less than a year and a half after becoming operational.

Now, Hempstead's second attempt at the ultimate answer to its garbage problem is rising on the ashes behind the facade of the

first plant. It is a $370 million project scheduled to go into opera-
tion in August, 1989, that, like the first one, is supposed to burn
garbage and generate electricity. Unlike the first plant, it uses a
technology that is said to have a 20-year successful history elsewhere in the world, and pollution controls that make it second to
none. And again, officials are voicing optimism.

"We think the conditions are such that it will be a model for
other plants on Long Island and possibly for New York State as
well," Harold Berger, regional director of the state Department of
Environmental Conservation, said upon issuing permits for the
plant in November, 1986.

But an examination of the process surrounding the award of the
contract for the second plant raises questions about how much
the town learned from its experiences with the first.

A *Newsday* investigation found that town officials awarded the
contract for the second plant after a hasty bidding process in
which the only competing bidder was eliminated on technical
grounds after less than a week of review. As was the case with the
first plant, neither the firm chosen for the second one, American
REF-FUEL, nor its parent companies, Browning-Ferris Industries
and Air Products and Chemicals Inc., has built a garbage incinerator.

Moreover, despite promises by then-Supervisor Alfonse D'Amato—now a U.S. senator—that town residents would not pay one
penny for the old plant, the price to build the new plant includes a
$35 million payment for the old one. In turn, American REF-
FUEL and Browning-Ferris intend to recover that amount—and
more—from town residents over 20 years as they recoup their
investment in the plant by burning Hempstead garbage.

All told, Browning-Ferris stands to make almost $1 billion in
profits from carting garbage off Long Island until the plant is
built, for burning garbage in the plant, and for carting unburned
garbage off Long Island. Meanwhile, the annual garbage bill for
the average Hempstead home is expected to rise to nearly $700 in
the next decade, more than four times the amount in 1987. The
only solace a town resident can take from that number is that if
the town continued to ship garbage out of town, the cost would be
even more.

Hempstead's experience dealing with garbage over the last two

The first Hempstead garbage-burning plant goes down in dynamite, March, 1987.

decades, political experts say, is an example of what occurs when municipal officials try to make quick decisions involving hundreds of millions of dollars in a crisis-charged atmosphere with limited information.

Indeed, because of what happened with the first plant, Hempstead has left a legacy for an entire industry to live down.

When officials of Ogden Martin Systems were faced in 1987 with making improvements to an incinerator in Massachusetts, the specter of Hempstead's failure loomed over their heads. "The simplest thing to do would be to knock it down and replace it. But then this would have been another failure, another Hempstead," said John Klett, Ogden Martin's vice president of operations. "The industry would have lost ground in the face of another visible failure."

The investigation into the second Hempstead plant found that:

• Browning-Ferris developed its billion-dollar garbage monopoly in Hempstead through three contracts awarded after reviews of a week or less, a time considered too short by some industry officials for a proper analysis of bids and shorter than the time the town has taken to award contracts for providing such items as spark plugs and paper. In one of those cases, Browning-Ferris was not the low bidder but won the contract after the town disqualified other, lower bids.

• The only official competitor for the incinerator contract, Signal Environmental Systems Inc., said it submitted its proposal as an accommodation to town officials but felt it had no chance to win the contract because of the influence of Browning-Ferris' attorney, former Assemblyman Armand D'Amato, brother of Alfonse D'Amato. As Hempstead supervisor, Alfonse D'Amato was the driving force behind the first plant.

• The town did not make a thorough check of whether Browning-Ferris had any legal problems before awarding the incinerator. The company was involved in legal difficulties at least as far back as 1980. Since 1984, the company or its officials have: pleaded guilty or been convicted of bid-rigging and price-fixing; paid a total of $15.7 million to settle court actions involving allegations that it bribed public officials, fixed prices, and rigged bids; been indicted by a federal grand jury on charges of deliberately pumping toxic wastes into a creek; been sued by the federal government for improperly disposing of toxic wastes elsewhere.

• Hempstead's contracts provide profit-making opportunities and income guarantees unusual for the industry. One provision allows American REF-FUEL to make a possible $124 million more over ten years from burning non-town garbage at the incinerator—and Browning-Ferris an additional $67 million from town residents, who will foot the bill for hauling away that ash.

In many ways, the Hempstead story is a microcosm of the garbage problem facing communities across the nation. It is a story of dealing with increasing amounts of garbage, decreasing amounts of landfill space, and efforts at short- and long-term solutions that include building incinerators and shipping garbage out of town.

What is unusual about the Hempstead situation is that, more than a decade ago, Hempstead's office-holders saw the looming problem and thought they had it licked.

But, instead, they would only make things worse.

The story begins in 1971, when then-Hempstead Presiding Supervisor Francis Purcell assigned then-Supervisor Alfonse D'Amato the task of trying to solve the town's garbage disposal problem. It was a problem of dealing with landfills that were filling up and incinerators that were aging.

After investigating other options, D'Amato found his solution after a one-day tour of a demonstration project in Franklin, Ohio. He decided the town should build a resource recovery plant that used an untested, complex technology by a firm, Parsons & Whittemore, that was more at home building and operating plants that processed trees into paper.

D'Amato's main selling point was that the company would take all the financial risk and residents would not pay for service until the plant was fully working. But in 1978, before the plant was completed, D'Amato, by then the town's presiding supervisor, amended the contract to have the town start payments to have some garbage processed.

The plant rarely worked as intended. There were mechanical problems and burst water pipes. Ash from the smokestacks sprinkled over the neighborhood. And it smelled. Area residents complained to officials from Hempstead to Washington. The town and the firm began wrangling over whether the company had met its contractual obligations.

Finally, the dispute, the mechanical problems, and the finances became too much. Parsons & Whittemore closed the plant on March 7, 1980, less than a year and a half after the plant was operational. Several weeks after the plant closed, the federal Environmental Protection Agency announced that dioxin, a highly toxic chemical, had been found in the plant's air emissions. But dioxin had been a surprise discovery. And the samples had been handled so much that the EPA could not determine how much dioxin was being emitted from the plant.

Instead of being back to square one, Hempstead had actually lost ground. And what once was merely a garbage problem was now becoming a crisis. In expectation that the plant would work, the town closed one of its two incinerators. Other incinerators—serving villages and the Five Towns area—also were closed. The garbage that had been sent to the original plant was now rerouted to the town's landfills in Merrick and Oceanside. Garbage was piled higher at both landfills and, in the case of Oceanside, wider.

By 1981, residents were angry. But Alfonse D'Amato was no longer around. He had been elected to the U.S. Senate. In his place was Thomas Gulotta, an assemblyman with a history of piling up huge victory margins in elections. Gulotta was no stranger to the Hempstead garbage problem. As an assemblyman, he had represented the district where the plant was located. "I was determined to take an insoluble problem and have in place the building blocks of a solution by the time I left," said Gulotta, who is now the Nassau County executive.

But it would be three years until the town would find the right blocks. During that time, Gulotta tried to determine if the existing plant could be rebuilt. He tried to negotiate a settlement with Parsons & Whittemore and the three insurance companies that had financed the plant. And, at first, things did not go very well. "They literally refused to bargain. They said, 'We're your only ballgame. We can outwait you,' " Gulotta said.

Rather than risk a lawsuit from the insurance companies by evicting Parsons & Whittemore from the plant site, he insisted that the builder of the next plant buy the existing one from Parsons & Whittemore and its financial backers, since the firm had defaulted on its bonds. This extra cost—it would work out to more than $35 million for the plant and turbines—would be included in the price of the new plant. "We set it up so the vendor would pay Parsons as much as they [Parsons] want but charge us the least in tipping fees," Gulotta said.

The project was partially financed with $288 million in tax-exempt industrial development bonds and $49.5 million put up by American REF-FUEL, which is to pay off the bonds and interest with the revenue it receives from the town in tipping fees and with electrical sales to LILCO.

Meanwhile, town residents produced more and more garbage, and the town ran out of places to put it. In 1984, Gulotta closed the Merrick landfill because the slopes were too steep for trucks to traverse and converted the closed incinerator there into a transfer station where 200,000 tons of garbage a year would be dumped, packed in larger trucks, and shipped almost 100 miles away to upstate Orange County.

To handle this task, the town turned to the second-largest garbage collection company in the nation, Browning-Ferris Industries. The firm's efforts in garbage began with the merger of Browning-Ferris, then a manufacturer, with two small garbage companies in the Southeast and expanded nationwide through an aggressive program of acquiring other companies.

Now publicly held with its stock traded on the New York Stock Exchange, the company has operations at 200 places in 41 states and 24 locations in Canada, Puerto Rico, Australia, England, Kuwait, Malaysia, Saudi Arabia, Spain, and Venezuela. The company employed 18,600 people at the end of its fiscal year in September, 1986. But it only had tried one venture on Long Island, and that one had left a bitter taste.

In 1983, just as the company was about to sign a contract to build the incinerator for the Multi-Town Solid Waste Authority, a consortium of western Suffolk County towns, the plan fell apart amid political bickering between officials in Huntington and Babylon and opposition by local garbage carters.

Company officials learned two lessons about business on Long Island from that experience, lessons that translated into actions later that year: They hired their first local attorney and made peace with local carters.

The lawyer they hired was then-Assemblyman Armand D'Amato, the brother of Alfonse. Armand D'Amato, in an interview, said he felt his brother's political status helped him get hired by

Browning-Ferris. Clifford Jessburger, president of American REF-FUEL and then a top Browning-Ferris official, would only say, when asked recently, "What lawyer would you hire?"

At any rate, D'Amato contended that Browning-Ferris stayed with him because of the quality of his legal work. Eventually, company officials would work out of Armand D'Amato's Mineola law office and become his firm's primary client. Meanwhile, D'Amato served briefly on a state commission trying to find solutions to solid waste problems.

To make peace with the carters, the company originally proposed calling them together for a meeting but later canceled the session after a conversation between Ed Crane, Browning-Ferris's vice president for sales, and Salvatore Avellino, head of the trade group representing Long Island carters and the man described by law enforcement sources as being in charge of the Long Island garbage interests of the organized crime family headed by Anthony Corallo. In the conversation, which was taped by the state Organized Crime Task Force, Crane assured Avellino that Browning-Ferris had no intention of trying to take over local collection routes, as it had elsewhere in the country.

Crane, in an interview in December, 1987, said that that he did nothing improper and added that his contacts with Avellino came after discussions with Browning-Ferris's lawyer.

The company gained its first Long Island foothold in 1984—and that foothold was in Hempstead, Armand D'Amato's home turf. Hempstead awarded Browning-Ferris the $67 million, five-year contract to haul town garbage off Long Island, even though the company's bid was the highest of three submitted.

The apparent low bid of $50.5 million by Cedarbrook Contracting Corp. was declared ineligible because it did not meet all of the requirements of the bid, including proof that the company had the necessary insurance and available capacity at a distant landfill.

In a subsequent letter to the town, Leon de Bremont, Cedarbrook's president, complained that the company could not meet all of the town's demands in the five days the town allowed. In an interview with *Newsday*, Hempstead's sanitation commissioner, James Heil, admitted that five days might have been too short a time for a bidder to comply with the bid requirements.

But Gulotta said the denial was based on Heil's assessment that there is no way Cedarbrook could come up with a backup landfill. "The chances of him [de Bremont] getting it didn't exist. He was not in the garbage-hauling business, didn't have the equipment or the resources to do it." De Bremont did not return telephone calls seeking comment.

The award of the garbage-hauling contract bought time for the town. Gulotta said it also convinced Parsons & Whittemore to drop its insistence that the old plant could be reopened and to negotiate its sale.

But it did not solve the town's garbage problem for the long run. So town officials began once again to think about a new incinerator, and, in the middle of 1984, the town hired CSI Resource Systems Inc., a Boston management consulting firm, to help it make plans.

Although CSI was supposed to provide engineering services as part of its contract, it is not an engineering firm, nor have CSI engineers who work on the Hempstead project obtained the necessary temporary licenses to practice engineering in New York, according to state records and a member of the state engineering board. State law requires that firms advising municipalities on procuring incinerators must have these credentials, according to Richard Kenyon, chairman of the state engineering board. "Not every aspect of that project is engineering," Kenyon said, "but the heart and soul of it is."

Gulotta said, "It was my understanding that CSI could reach out and obtain experience, including engineers. Other consulting firms were more narrow in scope. CSI had available to them financial consultants that could handle the fiscal considerations."

Heil said, "The engineering capabilities of CSI were adequate to address the broad technical issues. Hempstead officials have said they are satisfied with the firm's work."

But in making a crucial decision on the plant—how the town would pay for burning the garbage—the town did not ask CSI for a study. The town, after some consideration, chose a net tip fee system as the way town residents would pay for garbage disposal over the succeeding 20 years.

ASSET BECOMES A FINANCIAL LIABILITY

By Michelle Slatalla and Thomas J. Maier

The $180 million incinerator on the Hudson was sold as a technological marvel that would solve Westchester County's garbage problems and keep costs down by selling the electricity it generates.

But in its three years of operation, the plant has had its share of problems. Energy revenues have been less than predicted and have so far cost county residents an extra $30 million.

"Don't think it's cheap and don't believe people who say that it will be cheap," says Westchester County Executive Andrew O'Rourke. "There's a lot of aggravation to this, but it's the only game in town. So we have to do it."

Westchester's experience with its 2,250-ton-per-day garbage-burning, electricity-generating incinerator is one example of the different kinds of problems communities across the country are encountering with the new European incinerator technology.

County officials and the plant's owner, Wheelabrator Environmental System of New Hampshire, say the incinerator has lived up to basic expectations. "We've fulfilled our contract and burned the full amount of garbage, but, it was not trouble-free," says Plant Manager Bob Hensel. "Hopefully, we are achieving high reliability, and the plant is improving over every year."

It hasn't been easy. A *Newsday* examination of New York state Department of Environmental Conservation records confirms that the plant has had mechanical problems that caused a number of unscheduled shutdowns since it began burning garbage in October of 1984.

But the biggest problem for Westchester residents has been with expected revenues from the sale of electricity. Rather than partially off-setting the cost of burning the garbage, those sales have generated increasing net losses to the county each year of the plant's operation with no indication that the trend will be reversed. So far, those sales have cost the county almost $30 million.

This loss resulted from a miscalculation when county officials—in drawing up the plant's energy contract during the early 1980s—assumed that oil prices would continue their then-skyrocketing pace. In its deal with Consolidated Edison, county officials pegged the price of its plant power to the utility's electric rates—which are continually adjusted for the price of oil—rather than the state's guaranteed fixed rate of 6 cents per kilowatt-hour for waste-to-energy plants.

By doing this, officials hoped to make a better deal for the county. The catch, however, was the county's guarantee to pay Wheelabrator a minimum of 6 cents per kilowatt-hour, plus any cost-of-living increases. Thus, when oil prices dropped soon after the plant started up, the county had to pay Wheelabrator the guaranteed rate, while getting a far lower amount from Consolidated Edison. In 1985, the difference was $6 million, another $9 million in 1986, and approximately $14 million in 1987.

Some communities along Westchester's northern tier, who were expected to pay for the use of the county incinerator, were chilled by the rapidly rising electrical costs that they had to share and pulled out of their agreement. Instead, they are shipping garbage out of state. Meanwhile, county officials have been searching in vain for a loophole out of the plant's energy contract. "Thirty million dollars is a lot of money," says Edward Brady, chairman of the Westchester Board of Legislators. "Even when you add in the cost of building and operating the incinerator," Brady said, "it would have cost more than that to get rid of the garbage if we didn't have the plant."

Another looming problem that Westchester officials are just beginning to gauge is the cost of getting rid of garbage when there is a prolonged incinerator shutdown. A consultant's report in September, 1987, found that costs, in a worst-case scenario, could exceed $5 million to dispose of garbage without an incinerator for more than a month. Without a working landfill in the county, garbage would probably have to be shipped by truck or rail out of state, the study said. If the incinerator shuts down for more than a month, the study added, Westchester would have to declare an emergency and appeal for state aid.

Under a net tip fee system, the vendor assumes all the risks associated with a contract to sell electricity generated by the plant to Long Island Lighting Co., but also garners all the profits. The town's other option would have had Hempstead and the vendor sharing the energy risks and the profits under a predetermined formula. That would have provided higher first-year costs, but also the possibility—if energy prices rose faster than inflation—of lower costs to taxpayers in later years, and, as a result, overall.

Gulotta acknowledged that the town chose a net tip method without asking CSI for a study of which method would be cheaper, saying it was because the net tip fee met the town board's desire of a low first-year cost and a minimum of risk. "With a net tip fee we could reasonably anticipate the future," Gulotta said.

Another crucial decision involved the type of plant to be built, and by the summer of 1984, the town and its consultants were writing specifications for the new incinerator. The specifications called for a plant to be built at the site of the closed plant in East Garden City that would handle 2,250 tons of garbage per day, a major plant by industry standards. Competition would be restricted to firms that could provide what was considered proven European technology. The town also wanted a company with experience in the field. And the town set a condition that the company had to negotiate purchase of the failed plant from Parsons & Whittemore and the insurance firms.

That would prove to be a key factor in seeing to it that a Browning-Ferris affiliate closed the deal.

Based on the size and importance of the Hempstead project, the town could have expected proposals from three to seven firms, claims a Manhattan-based underwriter who specializes in selling incinerator bond issues. When Babylon asked for proposals on its incinerators, it received five; Islip got three; Huntington, four.

Yet of the more than a dozen companies that picked up bid documents from Hempstead, only two made proposals: American REF-FUEL, a subsidiary of Browning-Ferris; and Signal Environmental Systems (now Wheelabrator Environmental Systems).

Had it not been for a plea by Heil, according to a top Signal official, American REF-FUEL's would have been the only proposal.

Former New York Lt. Governor Alfred DelBello, president of Signal's incinerator subsidiary when it submitted its proposal for Hempstead, said, "We thought it was near impossible to win the contract." DelBello added that he was persuaded to submit a proposal by Heil. "The town wanted us to bid for the integrity of the process," DelBello said. "We were accommodating them."

"Signal was encouraged to respond," Heil said. "If he [DelBello] accommodated us, that's his reaction."

Two Signal representatives, who asked not to be identified, said they felt their company had no chance to win the contract to build the new incinerator because of the influence of Browning-Ferris's Long Island attorney, Armand D'Amato.

Gulotta denied that American REF-FUEL was favored and said that, if anything, D'Amato's presence caused the town to lean over backward against the firm.

"I understand the possible problems of perception," Gulotta said. "As far as the town was concerned, he [Armand D'Amato] was not an assemblyman but a lawyer representing a client. There is no restriction on a state legislator having a law practice. We had no right to dictate to [Browning-Ferris] who they would hire."

The proposals were due at 4 p.m., February 25, 1985, a Monday. The next Monday, Heil telephoned Signal to tell officials that they had lost. A number of industry experts told *Newsday* that a week is not enough time to evaluate such complex proposals. For example, New York City took six months to analyze incinerator proposals and pick a vendor to build a plant at the Brooklyn Navy Yard; Oyster Bay took more than four months; Huntington, five.

Dan Harkins of CSI, who helped Heil analyze the two proposals, defended a quick decision. "The fact that we can work at a fast pace is not germane to whether the results can stand up to scrutiny," he said.

The town has worked at a slower pace in other contracts, however. It took two months, for example, to award contracts in 1987 for spark plugs and swimming pool chemicals. The town spent more than three months choosing a supplier of office paper.

Harkins said that the two companies with incinerator proposals were evaluated in three fundamental areas: the buyout, price to the town, and experience.

American REF-FUEL had never built a plant. Browning-Ferris was a garbage hauling company. American REF-FUEL's other parent, Air Products and Chemicals, builds and owns industrial gas plants. It brought to the joint venture financial strength and experience in building industrial facilities. American REF-FUEL's president and chief engineer came from Browning-Ferris; its top financial officer from Air Products.

American REF-FUEL planned, however, to use a West German model that burned unsorted garbage and then allowed metals to be sifted from the ash. A company official said in an interview that the technology was easily adaptable to Hempstead's needs without major technical problems. The Hempstead plant, however, was to be three times the average size of the many plants around the world using the same technology, only six of which produced energy on a scale equal to what was envisioned for Hempstead.

Signal was running three plants and had just completed two more. Signal, Harkins said, clearly had more experience.

As for price, CSI favored American REF-FUEL's offer, which proposed a flat price for each year of the contract, over Signal's, which called for sharing energy profits with the town based on the plant's performance. If the plant worked as well as Signal said it could, CSI concluded, Signal's proposal was cheaper. But CSI also concluded that Signal's projection was overly optimistic. As a result, CSI said, American REF-FUEL's offer was the cheaper of the two. Floyd Hasselriis, a thermodynamics expert, however, said that CSI's analysis was based on too conservative an assumption about the energy potential of Hempstead's garbage and that Signal's contention about efficiency was reasonable.

On the third criterion, the purchase of the closed plant, American REF-FUEL presented a letter of intent from the insurance companies and Parsons & Whittemore agreeing to sell the old plant to American REF-FUEL for $30 million (the $5 million for the turbines would be added later). Signal merely had said that it would buy the plant; the price it would pay would not affect the amount of its bid.

Seemingly, that would have placed the two bids on equal footing. But such was not the case, for word had been passed that the insurance companies that provided the financing for the old plant did not believe a deal could be struck with Signal. The word was passed to the town by the attorney for the companies, George Farrell, a former assemblyman and former vice chairman of the Nassau Republican Party.

"George Farrell told us that [a lenders' agreement with Signal] wasn't likely to come," Gulotta said. He added in a later interview, "I decided the best way was not to become involved in endless negotiation, but to seek an amicable solution."

In an interview, Farrell said that negotiators for his clients met five times with Signal representatives and that no agreement was reached.

Gulotta said that he was faced with no choice but to pick REF-FUEL because Signal, in not having an agreement on the old plant, had not adhered to the terms of the request for proposals. But Bernard Melewski, counsel to the state Legislative Commission on Solid Waste Management, said the law does not bind a municipality to pick the firm farthest along in fulfilling a requirement in a request for proposals.

Hempstead officials said that before awarding the contract, they checked the references of American REF-FUEL and its major parent, Browning-Ferris. But John Pessala, the then-assistant town attorney who advised Gulotta on the development of the incinerator, said his check into Browning-Ferris did not venture into criminal charges. "I just spoke to the municipal authorities where Browning-Ferris had garbage collection projects," he said, noting that he reached 50 municipalities.

Pessala said he did not check the firm's legal history because he was asked to look only into the company's experiences in serving other municipalities. As a result, the town was not made aware of the raft of civil and criminal actions against the firm around the country, actions involving allegations of bid-rigging, price-fixing, bribing of public officials, and improper disposal of toxic waste.

Gulotta said he did become aware of some of the actions against Browning-Ferris but was told by the federal EPA and the state Department of Environmental Conservation that there was no reason he should not negotiate with American REF-FUEL.

After selecting American REF-FUEL, the town entered into negotiations with the company over a contract finalizing the deal. What emerged—a contract signed in November, 1986—was a document with provisions that are extremely lucrative for American REF-FUEL and its parent companies, including provisions that several engineers and a financial analyst consulted by *Newsday* said are unusual in the industry.

Overall, the contract allows American REF-FUEL a rate of return on its investment of 53 percent before taxes, considerably higher than the 30 percent standard in the industry. Over the 20 years covered by the contract, the company profit could be as high as $898 million, according to estimates based on assumptions company and town officials said were valid.

One of the specific provisions allows American REF-FUEL to solicit other customers for the plant, which has a capacity to burn 2,505 tons a day—11 percent larger than what Hempstead had sought. While those customers will pay for having their garbage burned—increasing American REF-FUEL's income—Hempstead taxpayers will foot the bill for trucking away the ash created by that burning. That is in addition to paying to burn its own garbage and to have its own ash trucked off Long Island.

A Gulotta aide could not recall what the public got in exchange for hauling away the additional ash. William Reynolds, REF-FUEL's top financial official, said the company insisted on the town hauling the out-of-town ash after Hempstead insisted on dictating to REF-FUEL where it could purchase out-of-town waste.

Town residents will also pay for the fuel oil needed to maintain furnace temperatures, which are critical to controlling air emissions. Heil contends this should not cost the town anything because the garbage picked up by the town should provide enough heat to maintain temperatures. Under most contracts, however, the operator of the plant picks up the cost, if any, of the oil.

In addition, while Hempstead taxpayers share almost none of the energy revenues, they will make up any differences if LILCO's payments are late, if LILCO is successful in overturning its energy agreement in court, or if LILCO goes bankrupt and a backup deal with Consolidated Edison is not as lucrative as the LILCO contract.

Moreover, what the public will pay for garbage disposal could climb dramatically if there is labor strife during construction, or if, at any time during the next 20 years, regulations change to make incineration more expensive. The largest risk—although the federal government is currently leaning against it—is the possibility that the federal government will declare ash hazardous waste. The extremely limited hazardous waste disposal capacity in New York State, experts say, would lead to a dramatic boost in the cost of garbage disposal.

No matter what its terms, signing the contract with American REF-FUEL meant that Hempstead had taken care of two pieces of its garbage puzzle. There was still one more piece to go.

This piece involved what would be done with garbage that could not be burned and garbage produced at times when the incinerator was not operating. According to town estimates, that could amount to 4.6 million tons of garbage by the turn of the century.

While the town worked on finalizing the incinerator contract, in August, 1986, it issued a request for proposals to truck that bypass garbage out of town. And once again, Browning-Ferris came out the winner.

The town received two bids, one for $910 million from Metropolitan Waste Inc., a partnership of two contracting companies, and one for $786 million from Browning-Ferris. No local truckers bid. That, experts said, is because the town required the posting of performance bonds so large that they could not compete.

As with the other two contracts, the town announced its award in less than a week. Harkins, of CSI, said the town had to move rapidly because this contract had to be in place at the same time as the $288 million in bonds for the incinerator were sold.

The final contract signed by the town and Browning-Ferris is significantly different from what truckers were asked to bid on. The town, after negotiations with Browning-Ferris, slashed the amount of raw garbage the company would need to haul, expressed in a weekly collection rate, from 13,500 tons in the request for proposals to 6,500 tons in the contract.

Kenneth Tully of Metropolitan Waste contends that, in effect, his company and Browning-Ferris bid on separate sets of specifications.

"If the number were 6,500 tons, there is no doubt in my mind our number would have been considerably less than its $910 million bid," Tully said in an interview. Heil said the difference in

Newsday/John Keating

Work continues on the new Hempstead incinerator.

the original prices between Browning-Ferris and Metropolitan was so great that the town felt it did not have to go back to Metropolitan for another bid.

Heil said he lowered the number of tons during negotiations with Browning-Ferris, who persuaded him that the lower figure was all the capacity the town would need. While cutting the amount of garbage involved, the town did not get a price cut.

Metropolitan Waste sued the town, seeking to overturn the

award. State Supreme Court Justice Eli Wager dismissed the case, ruling that the town had acted within the law in choosing Browning-Ferris.

Gulotta contends that there is a distinct advantage to giving Browning-Ferris a monopoly on garbage disposal in the town, because it creates clear responsibility for disposal. "There were arguments on both sides," Gulotta said. "The ultimate view was that it would be much smoother, and more cost effective, if one corporation handled the problem from beginning to end."

PART VI

GARBAGE AS GOLD

CHAPTER 12

THE EXPENSE OF EXPERTISE

By Walter Fee and Richard C. Firstman

With its old incinerator idle and its options running out, the city of Long Beach turned for advice in 1981 to a company of engineers specializing in the disposal of garbage.

"Long Beach was scared," recalled William C. Miller, Jr., a vice president of the consulting firm William F. Cosulich Associates. At the Oceanside landfill, where Long Beach was sending its trash, fees were rising and space was filling up. Pollution problems had closed the city's incinerator in 1979.

That summer, Long Beach hired a subsidiary of Miller's consulting firm, the Dvirka & Bartilucci division, and paid it $10,000. The firm recommended that the city sign a contract with a private company to remodel and operate the city's old incinerator.

Dvirka & Bartilucci brought in a team of companies to carry out the $13 million plan. One of them, Montenay Power Corp., then hired Dvirka & Bartilucci to design the new plant and supervise engineering on the site. As of December, 1987, Montenay had paid the consulting firm $600,000 for its services.

Under this arrangement, Dvirka & Bartilucci—the region's busiest garbage consulting firm—profited from a market it helped shape. First it was paid as a consultant to the city; then it was paid as a consulting engineer to the city contractor it helped bring in—a company that had previously worked with Dvirka & Bartilucci on a job in Glen Cove and later hired the firm for a major project in Dade County, Florida. The company said it did not have a conflict of interest in Long Beach because it was not representing both sides at the same time.

But the company's dual role shows how municipal officials, forced to make decisions about garbage disposal under crisis conditions, sometimes rely on advice from consultants who may have a vested interest in the outcome.

As federal and state governments have stepped back from their roles as advisers and regulators, it is the industry itself that has been guiding the decisions of municipalities. In becoming almost a permanent arm of government, consultants have helped convert garbage disposal from a government service into a private enterprise.

In the Long Island–New York City region alone, resource recovery engineers and lawyers have been paid more than $46 million since 1981 for advice and services to municipalities—most of it since 1984.

Long Beach City Manager Edwin Eaton, saying the city's garbage problem "just descended on us with a vengeance," said the city was forced to rely on the private waste-disposal industry. "You have to deal with the people who want to sell you systems, proven or not. And you have to deal with the 'Beltway Bandits,' as they call the consultants in Washington."

Islip Town Supervisor Frank Jones said, "They have an inordinate amount of influence. When you hire a consultant to do some road work, you have some feel for what you want, what it should be, where it's headed. In incineration, you have no idea what the state of the art is and you have to rely on them almost exclusively for judgments. I wouldn't know water wall or mass burn or RDF from Joe DiMaggio."

Drawn to such a lucrative and ever-expanding field, engineers, lawyers, and investment bankers are joining with major construction companies in making waste disposal a big business.

Companies from fields ranging from nuclear power to stadium food services have moved into the leadership of the incinerator-manufacturing industry.

And now, officials from all levels of government are joining the garbage industry, earning lucrative fees for advising their former colleagues or going to work directly for incinerator builders. Others are entering the investment banking field, where resource recovery plants have been a major source of lucrative fees—totaling nearly $200 million around the country since 1982.

In the last several years, New York State's lieutenant governor, New York City's sanitation commissioner, a former state environmental commissioner, two ranking Environmental Protection Agency officials, the head of a state commission on waste disposal, and a town supervisor and county legislator from Suffolk have all decided to become white-collar garbagemen.

Consultants have become such a big part of the world of garbage that, when western Suffolk's Multi-Town Solid Waste Management Authority closed down in 1983 without building an incinerator, its books revealed that it had spent $8 million for little more than advice and public relations.

These are only the most recent demonstrations of a fact of municipal life: Private enterprise and public works make a profitable combination—especially when politics plays a role. A *Newsday* computer tabulation found that most of the major commercial contributors to the Nassau County Republican Party over the last few years have been resource recovery contractors, ranging from engineers to electricians.

Because incineration has become their business, resource recovery engineers tend to inform and advise their clients in ways that promote the industry.

Miro Dvirka, a principal in Dvirka & Bartilucci, acknowledged that some consultants may have a subtle, "subconscious" bias that favors large-scale, high-tech resource recovery as a solution over other, less expensive long-term approaches like recycling. "It happens," he said. "They may be so wrapped up in their line of work that they're working with blinders."

James Barker, president of a major consulting firm, CSI Resource Systems, disputed the criticism that consultants have an inherent bias toward resource recovery plants. "People are ignor-

FIRMS AT THE TOP OF THE HEAP

By Bob Porterfield

Three companies with contracts to build incinerators on Long Island and in Brooklyn using imported European garbage-burning technology have emerged as among the leading incinerator builders in the United States.

Two of those companies—like many in the business—have moved into the field from the ill-fated nuclear power or federally subsidized synthetic fuels industries of the early 1980s.

Here is a look at the three companies:

• Wheelabrator Environmental Systems Inc.—Saugus, Massachusetts, a few miles south of Boston, is the birthplace of one of America's growth industries: large-scale, mass-burn garbage incineration. There, New Hampshire–based Wheelabrator Environmental Systems Inc. built the first large resource recovery plant using imported technology.

Once an active participant in the federal government's synthetic fuels program that collapsed in 1984, the company now hopes the resource recovery business will become a major source of profit.

Wheelabrator has undergone an extensive corporate metamorphosis, beginning as Wheelabrator Frye Inc., becoming Signal Environmental Systems Inc., a subsidiary of the Signal Companies, later being spun off as part of the Henley Group after Signal's merger with Allied Corp., and ultimately emerging in 1986 as a subsidiary of Wheelabrator Technologies Inc., a publicly traded company. It is the second-largest operator of garbage incinerators in the United States, with 14 plants worth nearly $2 billion in operation, under construction, or under contract. Wheelabrator is the American licensee for the Swiss Von Roll incinerator technology.

Wheelabrator has two projects in the New York metropolitan area—a $180 million plant operating in Peekskill in Westchester County that is owned jointly with the John Hancock Insurance Co. and a $445 million incinerator planned for the Brooklyn Navy Yard.

During the first six months of 1987, the Henley division that became Wheelabrator reported $12.2 million in net profit on nearly $500 million in revenues, much of that attributable to refuse incineration, and the company expects to increase its profits in this area.

Limited Experience

Four of the companies ranked as the top 10 in the garbage-incinerating business have not built any garbage plants that are currently in operation. Two others have only one plant in operation. This chart shows the number of units each company has been involved with and the number of tons of garbage per day the plants will process.

Company	Units operating No. Tons/day		Units under construction No. Tons/day		Units in contract No. Tons/day	
Wheelabrator Environmental Systems	7	12,200	3	3,325	4	6,850
Ogden Martin	3	2,050	7	7,812	7	11,050
American REF-FUEL	0	0	1	2,250	4	6,850
Combustion Engineering	0	0	3	8,000	1	750
Foster Wheeler	3	2,650	0	0	4	3,350
Westinghouse Electric	1	510	1	1,344	3	4,038
Dravo	0	0	2	1,880	2	1,890
Waste Management	1	1,000	0	0	1	2,200
Consumat	25	2,410	1	360	0	0
Thermo Electron	0	0	1	2,340	0	0

SOURCE: Newsday survey as of December 1987

• Ogden Martin Systems Inc.—
For years, Ogden Corp. has made
millions selling hot dogs and
beer at sporting events, refueling
aircraft, and providing a variety
of logistical services to industry.
But garbage disposal may
ultimately prove far more profitable, as the company quickly
becomes a leader in incineration
plants. From Long Island to the
rural slopes of Oregon's Will-
amette Valley, Ogden subsid-
iaries have 17 incinerators in
various stages of development or
operation representing a total
cost of about $2.3 billion—
including the $88.9 million, 750-
ton-per-day Babylon plant.
Ogden Martin Systems Inc.,
a Fairfield, New Jersey–based
subsidiary, entered the garbage
business in 1983 by obtaining
marketing rights for Germany's

Martin GmbH technology. A
second subsidiary, Ogden Allied
Services Inc., operates and maintains the facilities after construction.

Although 1986 revenues
from refuse-disposal operations
accounted for only 3.8 percent—
or $30.3 million—of the com-
pany's total revenues, garbage disposal accounted for nearly 25
percent of the company's $50.1
million operating profit.

• Combustion Engineering Inc.—
When Combustion Engineering
Inc. was awarded the contract to
build Huntington's $118 million
East Northport garbage incinerator in 1985, it underscored the
company's rapid entry into
refuse disposal. In less than
three years, the Windsor, Connecticut, firm won more than $1

ing the integrity of some of us professionals who work very hard
on the public's behalf—and ignoring in some ways the integrity
of the vendors."

Dvirka's firm—which has done work from Brooklyn to Sadat
City, Egypt—denies that there was any bias in the Long Beach
advice. "We did a very objective study for Long Beach," said
Miller, the firm's vice president. "The direction that Long Beach
took, they took relative to their own decision-making, not any
guidance or preconceived direction from D&B."

On Long Island, where towns have been forced into incineration by state mandates and a dearth of landfill space, the question
has become not whether to burn but how to do it. When they hire
consultants, municipal officials give the private sector great influence over decisions about what kind of plant they should build,
who should build it, and how big and expensive it should be.

"The system you choose, a policy decision, is done almost

billion worth of incinerator contracts, including one for the nation's largest plant—a 4,000-ton-per-day Detroit, Michigan, operation. Combustion Engineering attributes much of its success in this new field to its international reputation for design and construction of commercial and industrial power plants—a business that dwindled during the early 1980s. As a result, the nation's third-largest supplier of steam equipment for nuclear power plants was forced to diversify into petroleum exploration and production machinery, and later into synthetic fuels. The company's move into the garbage business began in 1984, at about the time the government abolished the synthetic fuels program. Today, Combustion Engineering supplies both the European-style mass-burn incinerators and the American-designed refuse-derived fuel plants.

While Combustion Engineering's refuse business is minuscule compared to its other operations, garbage incineration holds promise for huge future profits. In 1986, CE earned $50.8 million in net income on revenues of $2.55 billion.

Records show that only about 8 percent, or $197.8 million, of the company's total revenues were generated by its Public Sector and Environmental Division, but 93 percent, or $183.9 million of that, was attributed to garbage plants. Division operating profits totaled slightly over $4 million, and net profit was reported as "modest."

exclusively at the recommendation of the [consultant] you're dealing with," Frank Jones said.

"Now, all the knowledge is with engineers, consultants, lawyers, and bankers," said former Babylon Town Supervisor Anthony Noto, who has started a resource recovery consulting business with an assistant town attorney. Critics of the industry raise questions about the quality of the advice the consultants are giving.

On the advice of Dvirka & Bartilucci, the town of Huntington awarded a contract to a company, Combustion Engineering Inc., that has never built an incinerator. The company plans to build a 750-ton-per-day plant that will be the largest of its kind among 79 worldwide using the same kind of Italian technology. Some experts say this kind of upscaling often leads to problems. The $123 million plant, which will include a recycling center, will be one of the most expensive for its size in the nation.

"We hire consultants to explore it and question everything as we go along, both from the economic viewpoint and the disposal method," Huntington Supervisor John O'Neil said. "The experience [of the technology] was investigated by D&B, and it is a sound technology."

Dvirka said he carefully studied the technology of the planned Huntington plant and believes it will work. He noted that a report his firm did for the city of Palo Alto, California, recommended against high-tech incineration. "I blew $80,000 in fees," he said, referring to fees he could have gotten for consulting work on the incinerator.

On Long Island, consultants have enjoyed a big market for their services because they are considered the experts on a crisis virtually no municipality can avoid. And they have benefited from two key circumstances: the state's order that most landfills be closed by 1990 and the reluctance of municipalities to collaborate on regional solutions. Some observers are critical of such isolationist policies—and skeptical of the environmental studies being done by consultants.

"It's insane to have each town decide for themselves," said Lee Koppelman, executive director of the Long Island Regional Planning Board. "They are not expert in it. And the choice of consultants is often political. It's too complex and expensive an issue for this to be done in a business-as-usual way."

Bernd Franke, a Maryland-based consultant who has worked for such bodies as the Philadelphia City Council and who was hired by *Newsday* to evaluate the potential effects of proposed local incinerators, says there are flaws in a system that allows private consultants to exert great influence over the decisions of public officials.

"Every environmental impact statement goes through the same motions of describing the technology and estimating emission rates," Franke said. "But there really is no guide of how to do this, so every consultant can become rich by proposing studies using whatever data they want to use. A better way to spend millions of dollars would be to use it on a statewide basis to generate real criteria to estimate health risks, rather than to leave it up to individual consultants."

Consultants say that a lack of consistent state guidelines

means they are on their own in evaluating test results and design-
ing pollution-control equipment.

One of the reasons towns have been reluctant to work together
is the failure of the Multi-Town Authority. Multi-Town was supposed to build an incinerator to solve the garbage problems of
Islip, Babylon, and Huntington towns. But the state Commission
of Investigation reported in 1984 that the $8 million spent on
consultants and other services "has been lost irrevocably without
the towns' critical needs being advanced one iota."

Some towns have spent as much on their own—even before the
groundbreaking ceremony for a new incinerator. A prime example is the town of Oyster Bay, which has paid $9.5 million to
environmental lawyers and engineering consultants since 1981.
The bulk of the money—$7.7 million—has gone to the firm of
Lockwood, Kessler & Bartlett of Syosset. While $1.3 million of
that was related to the cleanup of hazardous wastes at the town's
Oyster Bay landfill, a federal Superfund site, most of it paid for
engineering and advice on resource recovery and landfill matters.

Oyster Bay also has paid $1.8 million to the law firm of Bev-
eridge & Diamond, whose founding partner is Henry Diamond, a
former commissioner of the state Department of Environmental
Conservation during the Rockefeller administration and a promi-
nent Republican. The firm also has been paid $485,000 for work
on the current Hempstead Town resource recovery plant.

The Cosulich firm, which includes Dvirka & Bartilucci, has
become the most visible member of the permanent government
of garbage in the Long Island–New York City region. Besides
Long Beach, it has advised Huntington Town ($2 million in bill-
ings), Glen Cove ($1.85 million), and the Brookhaven Town
($863,000). In New York City, it has split a $2.1 million contract
with another firm, Parsons Brinckerhoff Quade and Douglas Inc.,
a multinational construction consulting firm.

Town officials on Long Island say they have to hire consultants
because they cannot afford to keep their own staffs of experts.
"We can't hire engineers for three years and then fire them," said
John F. Delaney, executive director of the Solid Waste Management Authority of the town of North Hempstead.

But even in New York, where the city Sanitation Department's
$718 million annual budget includes a resource recovery office

From Government to Business

The garbage incineration business, like many multibillion-dollar industries that rely on government contracts for much of their income, attracts executives from the ranks of government officials. Here are some examples:

Robert LaBua

Served four years in the Suffolk County Legislature, where he was a member of the county's solid waste task force and the Long Island Regional Ashfill Board. In January, 1986, less than a month after leaving the legislature, he signed a $55,000 contract with Suffolk as a consultant on solid waste management. The county has paid LaBua's company, Construction Techniques of East Northport, $45,400 since February, 1986.

Alfred DelBello

As Westchester County executive, DelBello approved a contract to build a $180-million incinerator to be built by Wheelabrator Technologies Inc. Later, while lieutenant governor, he promoted municipal use of incinerators. In 1985, he quit to join a company that later merged into Wheelabrator, which is planning an incinerator in Brooklyn, and wound up as vice president for marketing. A special Westchester investigator later ruled that DelBello had not violated any ethics codes. Last year, DelBello earned $274,583 in salary and bonuses from Wheelabrator.

Gordon Boyd

Executive director for the State Legislative Commision on Solid Waste Management since it began in 1984. Helped make policy that has spotlighted the landfill crisis and promoted a move to incinerators. Leaving this month to form a consulting business with two other state staffers. Says they will help environmental-services and waste-management firms "tailor their business approach to the needs of the government." Says he sees no ethical conflicts.

Norman Steisel

As New York City sanitation commissioner from 1979-86, approved contract to build a $445 million incinerator at the Brooklyn Navy Yard. About three weeks after that action, he resigned to go to work for Lazard Freres, one of the investment banking firms involved in financing the plant. A special investigatory panel cleared Steisel of any improprieties.

David Sussman

A former Environmental Protection Agency official who strongly supported the use of European-style incineration technology. Now a vice president of Ogden Martin, which plans to build 14 plants nationwide using the European-style methods, for a total pricetag of $2.3 billion.

Henry Diamond

As the state's first environmental conservation commissioner, he pushed in the early 1970s for passage of the $1.15-billion Environmental Quality Bond Act, which has funded several incinerator projects. Later quit state post and joined a Washington-based law firm. During the 1980s, Diamond's firm, Beveridge and Diamond, was paid a total of $2.3 million by Hempstead and Oyster Bay to oversee the legal work on their incinerator projects.

Anthony Noto

Former presiding officer of the Suffolk County Legislature; lost bid last month for re-election as Babylon supervisor. After losing, Noto decided to open On Line Management, a consulting firm specializing in solid waste. His partner will be Leonard Shore, an assistant town attorney in charge of Babylon's resource-recovery plant.

Garrett Smith

Left post with the air and waste management division of the Environmental Protection Agency's New York office to work as recycling coordinator in Middlesex, N.J., and later as solid waste director in Essex County, N.J. Joined Ogden Martin Systems, Inc., one of the country's biggest builders of waste-to-energy incinerators, as director of recycling earlier this year. "I feel I'm likely to make a more substantial contribution here at Ogden to solve the garbage problem than if I stayed in government."

staffed by 81 people, consultants have played a major role. Since 1981, the city has spent nearly $19 million for outside help.

Engineers and lawyers are not the only white-collar garbage people profiting from the trend toward incineration. For every plant, there is an investment banker. For underwriting $13.5 billion in bonds used to finance plants since 1982, Wall Street's investment bankers have earned $194 million in fees.

In the case of the plant proposed for the Brooklyn Navy Yard— the first of five New York City may build—the investment firms of Merrill Lynch, Goldman Sachs, and Lazard Freres stand to net at least $14 million for underwriting $398 million in industrial revenue bonds.

Whether they are engineers, lawyers, or investment bankers, critics say most municipal waste disposal professionals are in the same business: selling incineration to the exclusion of other options.

"The consulting engineers don't really make money by saying to a community, 'You can solve your garbage problems once and for all by drastically reducing the amount of garbage you produce,' " said Walter Hang, who heads an environmental unit of the New York Public Interest Research Group. "Consulting firms make money when they say, 'We'll design an incinerator, we'll help you finance it, we'll help you build it, we'll help you operate it.' "

CSI Resource Systems Inc., a national consulting firm based in Boston, is one such full-service company. It has evaluated resource recovery for agencies ranging from the federal Environmental Protection Agency to the U.S. Conference of Mayors. It has earned $1.8 million in fees from the town of Hempstead and another $680,000 from New York City. But CSI does not work exclusively for public agencies. Its president said recently that the firm is now also in the business of helping incinerator firms gain operating permits.

Although the company presents itself as an engineering consulting firm, it is not licensed to practice engineering in New York State. State law requires that firms advising municipalities about incinerators must have engineering licenses, according to Richard Kenyon, chairman of the state engineering board. Company officials say engineers on staff are licensed in other states.

INCINERATORS HOT ON WALL STREET

By Bob Porterfield

Since 1982, Wall Street invest-ment bankers have reaped about $194 million in fees for raising the money to build garbage incinera-tors. In all, the bankers have floated tax-exempt bonds worth $13.5 billion.

Those underwriting fees also have attracted the attention of in-stitutional investors such as in-surance and automobile credit companies who are beginning to make inroads into the financing of incinerator construction.

Before the Tax Reform Act of 1986, the tax laws were extremely charitable to private garbage-plant operators, giving them investment tax credits, energy tax credits, and accelerated depreciation. With the demise of these benefits, the trend is away from private ownership of the plants. Instead, governments are becoming the plant owners but are leasing them to private firms that operate them. Financial ex-perts say this is because incinera-tor bonds are still tax-exempt under the law as long as plants are publicly owned. Future bond is-sues will be larger—and thus more costly for taxpayers—be-cause of reduced contributions by private industry toward the capi-tal costs of the plants.

Robert E. Randol, of Smith Bar-ney, Harris Upham & Co.'s munic-ipal department, predicts growing numbers of new municipal bond issues for garbage incinerators. He says that garbage will still provide plenty of opportunity for private sector investment.

The growing volume of institu-tional investment in garbage is being fueled by large financial ser-vice companies such as John Hancock Mutual Life Insurance Co., Ford Motor Credit Corp., General Electric Credit Corp., and others who see incinerators offer-ing huge, long-term potential for capital investment.

John Hancock was one of the first to invest in resource recovery, forming a partnership with what is now Wheelabrator Environmen-tal Systems Inc. to own the West-chester County incinerator in Peekskill. Hancock prefers eq-uity—or partnership—interests but also finances the debt incurred during plant construction, accord-ing to Herbert Magid, senior in-vestment officer in Boston-based Hancock's Bond and Corporate Fi-nance Department. More recently, Hancock was the lead lender in Wheelabrator's Millbury, Massa-chusetts, plant, along with several other investors, including Ford Motor Credit Corp.

Although Hancock and other in-surance companies will not reveal the rate of return on their incinera-tor investments, publicly reported revenue projections for Peekskill indicate that the partners who invested $51.7 million expect to receive revenues totaling $894.7 million over the next 16 years.

Robert E. Chambers, manager of municipal finance for the Ford Motor Credit Corp., agrees that waste disposal projects are lucrative investments. Garbage disposal has already reached "a crisis situation" in many parts of the East, and the problem will continue to grow nationally, he said. "If we felt it was going to go away, we'd find somewhere else to put our money."

As new tax laws discourage private investment, refuse plants will become more costly, and local governments will have to find alternative means of financing, Chambers said. Private investors, like the Ford Motor Credit Corp., are one of those alternatives, he said, and they will continue to invest as long as the return is acceptable.

Randol, whose department is the leader in municipal refuse issues with a volume since 1981 totaling $2.9 billion, said the explosion in underwriting activity during 1984–1986 occurred as deals were accelerated to beat tax deadlines—and exhausted the backlog of project financing. As more plants reach the financing stage, activity should pick up.

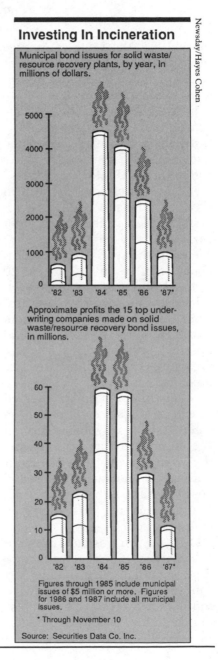

Newsday/Hayes Cohen

Investing In Incineration

Municipal bond issues for solid waste/resource recovery plants, by year, in millions of dollars.

Approximate profits the 15 top underwriting companies made on solid waste/resource recovery bond issues, in millions.

Figures through 1985 include municipal issues of $5 million or more. Figures for 1986 and 1987 include all municipal issues.

* Through November 10

Source: Securities Data Co. Inc.

ROLE OF A LOCAL ATTORNEY

By Thomas J. Maier

Since 1983, the issue of garbage disposal has played a significant role in Armand D'Amato's professional and public life.

In 1983, D'Amato, then an assemblyman from Baldwin, was hired by Browning-Ferris Industries as its local counsel. D'Amato, the brother of U.S. Senator Alfonse D'Amato, represented the firm in its negotiations to build an incinerator in Hempstead and in other garbage-related negotiations with the town that will bring the firm about $1 billion over the next 20 years. He has also sought garbage-related business for his law firm from officials in Huntington, North Hempstead, and in several upstate communities where Browning-Ferris wanted to site landfills.

Meanwhile, D'Amato served briefly on a state commission on solid waste management and, on at least one occasion after leaving the commission, talked about both private garbage ventures and possible state garbage-related action in a meeting with a Long Island town official.

D'Amato, who resigned from the Assembly in 1987 to devote more time to his law firm, said he has done nothing that constituted a conflict of interest. "The lay person doesn't appreciate and understand the conflicts of interest," he said. "There's an appearance and then there's the real thing. I have not done anything different than any other state legislator with a legal practice."

But two top state legislators disagreed and said they would have not allowed him to become a founding member of the state Legislative Commission on Solid Waste Management had they known about his relationship with Browning-Ferris.

"I probably would not have appointed him if I had known," said Assembly Minority Leader Clarence Rappleyea (R–Norwich), who named D'Amato to the panel in July, 1984. "It would certainly be subject to question."

The commission chairman, Assemblyman Maurice Hinchey (D–Saugerties), said D'Amato's serving at the same time he was a Browning-Ferris lawyer was a violation of the state's code of ethics.

Hinchey said his staff learned of D'Amato's role with Browning-Ferris late in 1984, while working on a report about organized crime and the waste industry. The state report cited Browning-Ferris's history of civil and criminal penalties and mentioned D'Amato's ties to the company. "I felt it was a conflict," Hinchey said. "He could influence public policy in . . . that he was employed by the nation's second-largest waste management firm."

D'Amato said he joined the commission because Long Island faced a garbage crisis and he hoped to help find solutions. But he also said that he never attended any of the commission's meetings while he was a member. "If you look at it in context, I really think it's a non-issue," D'Amato said. "I know it sounds good—solid waste commissioner represents a company involved in solid waste. I'll grant

Armand D'Amato and the incinerator.

you that. But the fact of the matter is that I never attended any meetings. And number two, I wasn't on it for very long. And number three, there were never any formal votes taken."

He said he left the commission in February, 1985, on his own initiative as his activities for Browning-Ferris increased. "I'm proud I got off of it on my own," D'Amato said, noting that he had concluded in his own mind "that there might be an appearance of a conflict, even though there was none."

Hinchey and Gordon Boyd, the outgoing executive director of the commission, supported D'Amato's account that he did not show up for meetings and had little impact on the commission.

A question about D'Amato's dual roles was raised by Islip Town Attorney Guy Germano. He recalled a meeting in early 1986, set up by Ross Patten, a top Browning-Ferris regional official, to talk about possible garbage ventures. D'Amato accompanied Patten to the meeting.

"The whole thing was bizarre," Germano recalled.

At first, the conversation centered on whether Islip, which was building one incinerator and considering a second, would consider hiring Browning-Ferris as the contractor for the second plant and to set up a transfer station to haul away Islip's garbage. Germano said Islip was not interested.

Then, Germano said, the con-

In a February, 1985, report prepared for the U.S. Energy Department, CSI described mass incineration as a "proven, effective method" of burning waste and included in a list of vendors Browning-Ferris Industries—a firm that had never built a plant. Later, CSI advised one of its clients, Hempstead Town, to pick a subsidiary of Browning-Ferris, American REF-FUEL, as the builder and operator of its planned incinerator, which is now under construction.

Barker, president of CSI, said in an interview that the list of companies in the federal report was not intended as a recommendation. "That sort of list is simply to say, 'Here are the guys with whom we have negotiated contracts.' It doesn't say those are the best guys.

"The technology that they're employing is proven," he said. "To the extent that there are people in the press that take issue with that, so be it. From an engineering point of view, the technology being employed by BFI or American REF-FUEL—whatever you

versation turned to an upstate landfill that Browning-Ferris wanted to build, with D'Amato suggesting that Islip push for state assistance for the company. Germano says he was offended by D'Amato's willingness to blend his role as a public official and private lawyer.

"I didn't throw him bodily out of the office, but I was annoyed," Germano said. "Do I think it was inappropriate for BFI to set up a meeting and then have a state assemblyman come in and talk about possible state legislation? The answer is yes."

D'Amato said he was not trying to influence legislation unduly, but rather to convince Islip officials that an upstate landfill was in the town's interest. The company has yet to gain approval for the landfill.

Armand D'Amato acknowledges that his brother's political status helped get him hired by Browning-Ferris. "I won't disagree that that was a factor," D'Amato said. "But talk to any successful developer, and they will say that [clients] stay with a law firm because they are good and get results. That's the bottom line."

When asked why D'Amato was chosen, Clifford Jessburger, then a top Browning-Ferris official and now president of an affiliate, American REF-FUEL, said, "What lawyer would you hire?"

wish to call them—is a proven technology." Other experts have said that such technology has not yet been proven effective in this country.

While governments buy most of their goods and services through a process of competitive bidding, they are permitted to hire consultants on the basis of much more subjective criteria. By its nature, it is often in some measure a political process.

Sometimes, political contributions and government contracts cross paths. According to a computer tabulation based on Nassau County Board of Elections records from 1981 to 1987, seven of the ten most generous commercial contributors to the Nassau Republican organization have been waste-disposal consultants or other companies with garbage-related contracts in the county. In all, such companies have contributed more than $75,000 since 1981.

The study showed that, aside from local Republican committees, the leading contributor to the party and its top elected official, County Executive and former Hempstead Presiding Supervisor Thomas Gulotta, was a firm whose public contracts include major resource recovery projects on Long Island.

The top contributor between 1981 and 1987, at $10,175, was Charles R. Velzy Associates, a prominent Carle Place–based engineering firm that has worked for local governments on a wide variety of projects. In the same period, the town of Hempstead paid the firm more than $1 million for consulting work on its second resource recovery plant. The firm has also worked for the towns of Islip and Smithtown.

The list of the top ten contributors also included five firms with contracts for concrete, electrical work, and other construction related to resource recovery plants. And sometimes municipalities and firms hire lawyers who have the kind of political clout that can ease the process of opening waste-disposal plants.

While he was a state assemblyman several years ago, Armand D'Amato went to work as a lawyer for Browning-Ferris Industries, the parent company of the firm building Hempstead's second incinerator. While in BFI's employ, D'Amato became a founding member of the State Legislative Commission on Solid Waste Management.

The town of Huntington, dominated by Republicans, hired Shea & Gould, a politically powerful law firm with strong Democratic connections, to help with its incinerator plans.

Asked if town officials had hired the firm in the hope it could help smooth relations with a state environmental agency that operates under a Democratic administration, town spokeswoman Pat Reiss said it "certainly wouldn't hurt." She added, "You use the people you think can help you the most."

Some of those people, like D'Amato, have jumped to the industry from the government side.

David Sussman, a former official of the EPA, took a job with incinerator manufacturer Ogden Martin Systems and has become one of the industry's leading advocates.

Gordon Boyd, formerly executive director of the State Legislative Commission on Solid Waste Management, left his public job to become a consultant. He says that environmental companies have legitimate needs for people who know how government works. "Are we indentured to the original job we had?" he said in

Advice Is Not Free

Amounts paid by Long
Island cities and towns to
consultants who worked on
landfill and resource-recovery
issues

Town/City	Total
Oyster Bay	$9,526,000
North Hempstead	4,434,246
Hempstead	3,652,612
Huntington	3,342,375
Babylon	2,587,954
Glen Cove	1,855,216
Islip	1,158,446
Brookhaven	1,136,565
Smithtown	101,696
Long Beach	10,000
Long Island total	27,805,110
New York City	18,789,248
Area total	46,594,358
SOURCE: Town and city records	

Newsday

an interview. "That doesn't make much sense. I don't feel any
reason to feel defensive about it."

And after losing his bid for re-election, Babylon Supervisor
Noto decided to open a consulting firm with Leonard Shore, an
assistant town attorney in charge of the town's resource recovery
plant. Noto envisions his firm as a kind of production coordinator. "We can be handholders, up front with the resource recovery
team, through the process of RFP [request for proposals], picking
engineers and lawyers," Noto said. "We know where the pitfalls
and potholes are."

THE FIRM'S DAYS IN COURT

By Robert E. Kessler

Browning-Ferris Industries has a considerable history of being hauled into court to defend itself against charges of price-fixing, bid-rigging, bribing public officials, and improperly disposing of hazardous waste.

As a result of various court actions around the country, the company has already paid $15.7 million and faces penalties of millions more in cases ranging from civil suits in Ohio, Pennsylvania, and Vermont, to criminal actions in Ohio and Louisiana.

In addition, in December, 1987, five separate federal grand juries investigated charges that the firm fixed prices in Phoenix, Pittsburgh, Cleveland, Memphis, Tennessee, and Birmingham, Alabama, according to federal and company officials.

Andrew Maloney, the U.S. attorney for the Eastern District of New York, is investigating the firm as part of a probe of the waste collection industry on Long Island and in New York City, according to sources. Maloney declined to comment, except to say, "We are aware of BFI's checkered history."*

Company officials say the firm is law abiding and has a vigorous

* No indictments have resulted from these investigations to date.

policy of insisting that its officials follow the law, including the anti-trust laws. "Considering the size of the company and the number of operating locations ... I don't think it is a large number," company attorney John Potwin said of the court cases and investigations, noting that the firm has 18,000 employees in 200 locations.

But Assistant U.S. Attorney General Charles Rule, who as head of the Justice Department's Anti-Trust Division is in charge of the grand juries investigating Browning-Ferris, said, "It's pretty easy to be excluded [from indictments] by not violating the law. . . . They keep committing violations. It certainly raises questions, doesn't it?"

The cases involving the company include:

• A guilty plea in November, 1987, to criminal anti-trust charges in Toledo, Ohio, and an agreement to pay the maximum penalty, a $1 million fine for rigging bids and fixing prices on garbage collection contracts. Company officials say the case resulted from a local manager violating firm policy. Browning-Ferris paid the state of Ohio $350,000 to settle a parallel civil case.

- An ongoing civil case in Pittsburgh involving accusations by a rival firm that Browning-Ferris secretly subsidized small carting firms to make it appear as if there were competition and disposed of hazardous waste in landfills that were not designed to accept the substances safely.

  In July, 1987, U.S. District Court Judge Gerald Weber withdrew from the 4-year-old case, saying the actions of the company and its attorneys in court proceedings could have so prejudiced him that he might not be able to render impartial rulings. "There is no doubt that BFI ... misrepresented to the court and to other parties that certain [corporate] documents did not exist when in ' fact they did exist. There is also no doubt, by the defendant's own admission, that [Browning-Ferris] ... destroyed documents which they were court-ordered to produce," Weber said before withdrawing.

  Browning-Ferris officials said the firm had done no wrong and that the judge should have withdrawn because he improperly met with attorneys for the rival firm.
- A jury verdict in April, 1986, ordering Browning-Ferris to pay $6 million to a small garbage carter in the Burlington, Vermont, area. Company tactics, according to court records in the case filed in June, 1984, involved plans to slash prices to drive the rival out of business, and after that to double prices. A company vice president, according to court records, said of the rival firm: "Squash him like a bug. . . . Put him out of business; croak him." Browning-Ferris is appealing.
- A lawsuit filed in Louisiana in April, 1987, charging the company with failing to dispose of wastes properly at its toxic waste site in Livingston. Fines in the case, if the federal government is successful, could run up to $2 billion.*
- A pending criminal trial of a June, 1985, indictment charging the company and two of its officials with deliberately pumping contaminated water into a creek near the firm's toxic waste dump outside of Cincinnati. The company denies any wrongdoing.
- Two anti-trust cases in Atlanta that resulted in $480,000 in fines against the company and, in one

---

* The case was settled out of court with Browning-Ferris Industries agreeing to pay $2.5 million in penalties.

of the cases, a 45-day prison sentence for the company manager in the area. Despite pleading no contest to conspiring to fix prices, the manager rejoined Browning-Ferris after serving his sentence. The manager "had done his time and paid his dues," said Potwin, the company's lawyer. "[He should] not be punished further."

One of the cases, in 1985, charged the firm with price-fixing in connection with the contracts to remove garbage from government buildings in the Atlanta area. That case resulted in the only time the firm has been barred from bidding on contracts, according to Potwin.

- A state anti-trust case in New Jersey in which the company agreed to pay $3 million and a then-company vice president pleaded guilty to an anti-trust charge and paid a $50,000 fine. The 1984 case alleged that the vice president bribed two public officials to obtain municipal contracts. One of the New Jersey officials pleaded guilty to official misconduct. The other died before being brought to trial.

- A civil anti-trust suit in Houston in which the company, in a settlement, paid $5.2 million in 1984 after being accused of paying a $25,000 bribe to a Texas state senator to block state approval of a landfill permit for a competitor. At the time, Browning-Ferris owned and operated the only landfills in the Houston area. "There are all kinds of reasons why a case is settled," Potwin said, noting that no company officials have ever been charged criminally with giving a bribe in connection with the Texas case.

PART VII

SOLUTIONS?

CHAPTER 13

THE SEARCH FOR SOLUTIONS

By Ford Fessenden

No notion in the field of garbage is so widely praised and so much ignored as the simple idea of not producing so much in the first place.

"It's just like the energy crunch, where we found that the best way to get through it was to not use so much in the first place," said David Anderson, a member of Connecticut's Solid Waste Advisory Council. "The cheapest way to solve the garbage crisis is to not make so much of it."

But virtually every attempt to reduce packaging waste, which has become the largest component of garbage, has met with failure. Since 1960, the amount of packaging in garbage has increased 80 percent. It is now a third of what Americans throw away.

In the last decade, New Jersey has considered a ban on clamshell-style plastic foam boxes used for fast food; Oregon has pondered a ban on disposable diapers and plastic grocery bags; Rhode Island's legislature debated a plastic-waste disposal act; Vermont has considered bottle-type deposits on tires and batteries. Several states have considered surcharges on difficult-to-dispose-of items: Ohio proposed a $2.25 surcharge on tires, and New York and Minnesota have considered similar measures. None was implemented.

"There are a lot of good ideas," said Elliott Zimmerman of the Illinois Department of Energy and Natural Resources, who has completed a study of source-reduction initiatives, concluding that a "lack of political will" was largely responsible for their failures. "The industry fights it, and they've been very effective."

A whole branch of American produce-sell-consume economics has evolved around the package. Great amounts of creative thought and research aim at coming up with packages that do not just contain a product, but also sell it. There is virtually no such time committed to the increasingly vexing question of what happens to the package after that.

Thus, someone like Frank Dittman, the associate director of the Plastics Recycling Research Center in New Jersey, which is trying to develop a way to recycle plastic bottles, has to spend so much money trying to get the aluminum caps out of his ground-up plastic that a potentially lucrative recycling strategy ends up being far less profitable. Dittman said bottlers want to be able to choose between plastic and aluminum caps, depending on price and availability.

"The best thing is for the manufacturers to stop using them, but manufacturers don't want to for economic reasons," he said. "They want to be able to play one supplier against the other."

There are some signs that the garbage crunch may begin to change that.

"There is no choice but to begin to look at packaging," said Richard Kessel, executive director of the New York State Consumer Protection Board, who has asked McDonald's to stop using polystyrene boxes for hamburgers. Mayor Edward Koch has said that he was joining that effort. "Unless we direct our attention to various products," Kessel said, "we are going to be buried in our garbage. We're going to be wrapped up in packaging."

McDonald's contends that its plastic foam amounts to a fraction of one percent of the nation's garbage, and to stop using it would not solve the problem. "We are absolutely looking and working [for alternatives]," said McDonald's spokeswoman Lana Ehrsam. "But for now, foam packaging is important to us."

Compacted soda bottles ready for shredding and recycling at New Jersey's Rutgers University.

Ruth Lampi, executive director of the Environmental Task Force, a Washington-based environmental group that helped to stop the marketing of a new plastic can, believes the scattered efforts of the states have acted as "consciousness raisers," and are the prelude to initiatives that will be undertaken more seriously, perhaps at the national level.

"If everybody had a wish list, the wish would be to reduce the amount at the source," said Jan Ray Clark, environmental specialist with the Solid Waste section of Florida's environmental agency. "But I don't think any state alone can do that. That seems to me to be a national initiative."

Congressional hearings during the summer, 1986, on the garbage problem raised the waste-reduction issue seriously for the first time since bottlers and packagers helped to defeat a national bottle bill in the 1970s.

"There's a role for some kind of fees to provide incentives to cut down on materials that are more deleterious [to the environment] than others," said Senator Max Baucus (D–Mont.), chairman of the Senate Environment and Public Works Committee.

"It's that kind of question that will begin to be asked," said Ron Cooper, an aide to Baucus. "We have to ask ourselves, how do we make this in a fashion that it doesn't pollute? How do we make it so that it can be recycled? Why do we continue to increase this through packaging? We've never really separated it down and looked at the problems."

The Islip garbage barge held important lessons about our refuse: Increasingly, it is made up of things that do not smell and of things that can or should be recycled.

"I saw pictures of the barge close up, and it made me angry," said Rod Edwards, head of recycling programs for the American Paper Institute. "It looked to be a high percentage of corrugated boxes. It occurred to me they were just too lazy to separate it, and there's a critical shortage of that kind of wastepaper."

Recycling company officials say the fastest-growing materials in the garbage heap over the last 25 years are those for which the imperative to recycle is the strongest: Paper and aluminum—the most readily marketable materials—and plastics, which are not easily marketable, but should be removed from household trash because they do not break down in a landfill and can corrode incinerators.

Recycling's environmental and monetary appeal is compelling. Reuse of material is much more environmentally benign than its principal alternatives, landfilling and incinerating, and it is usually less costly, studies have found.

For instance, a pound of paper burned in an incinerator generates about 500 BTUs of steam, according to Donald Walter, director of biofuels and municipal waste technologies at the U.S. Energy Department. Recycling that same pound conserves about 2,000 BTUs in energy required to produce paper from virgin pulp.

In the era of skyrocketing tipping fees at landfills and incinerators, recycling becomes more attractive as a way to avoid costs. A study by the California Waste Management Board set the statewide average cost of collecting and disposing of trash at $60 a ton. But the net cost of collecting and recycling was about $40 a ton, the study found.

Similarly, a study by a public policy research group in Minnesota found that recycling cost a third of what burying and burning cost in Minneapolis and St. Paul.

The Cuomo administration has proposed a number of strategies for encouraging recycling:

- Mandatory separation of recyclable materials in the home, in the manner that New Jersey has pioneered.
- Tax credits to lure manufacturers that use recycled materials, so that markets for the collected bottles, papers, and cans can be improved.
- A state brokerage to facilitate sale of materials, and state money for recycling equipment and demonstration projects.
- State standards and demonstration projects for composting organic wastes like leaves and food.

"The reason you go to source separation is to begin to create a market for materials," Frank Murray, Cuomo's environmental adviser, said. "But that alone won't solve the problem. You have to focus on the private side. The mind-set that we have is that a manufacturing industry in this state is based on raw and virgin materials. We're going to try our best to change that."

And on the federal level as well, there are signs that the Environmental Protection Agency, spurred by the barge, will begin at least to look at similar measures. On October 9, 1987, EPA officials called a meeting with state recycling officials.

"Part of the meeting, quite frankly, was to bring EPA up to speed on what was going on in the recycling world," said Richard Keller, manager of recycling programs for the Maryland Department of Energy. There was "a good bit of discussion as to what role EPA should be playing in this whole area, and how they could become more involved in market development activities, get more involved in regional programs," Keller said. "EPA hasn't shown this level of interest since before the Reagan administration."

There are questions about the cost effectiveness and environmental effects of burning. But it also seems clear, even to many environmentalists, that incineration of garbage must form some part of the answer to the garbage problem.

"A sizable portion of the waste stream, 30 to 40 percent, is not

OFFICIALS TAKE ON TAKEOUT CONTAINERS

By Thomas J. Maier

Thirty years ago, a customer looking for a quick hamburger would probably find it at a diner, served up on a ceramic plate dulled by frequent use.

But today's typical hamburger is passed over the counter of a fast-food outlet, encased in a plastic foam carton designed to be used only once.

Eventually, most of these clam-shell-like boxes and other plastic foam packages—more than 4 billion pounds a year, according to one industry estimate—wind up at the local dump or incinerator.

To stem the tide of fast-food garbage, several state and local officials recently proposed bans on the use of the plastic foam packaging commonly referred to as Styrofoam, but more accurately called polystyrene.

In New York City, Mayor Edward Koch followed state consumer officials in asking that McDonald's and other fast-food vendors stop selling their products in the plastic foam containers. In April, 1988, the Suffolk County legislature signed a ban on the use of the packaging. And communities in Vermont, Rhode Island, and California have considered similar restrictions.

Critics say the nonbiodegradable plastic foam is an unnecessary waste of limited landfill space and creates dangerous fumes when incinerated. "Burning the Styrofoam in the incinerators would make deadly poisons, like dioxin," said legislator Steve Englebright (D–Setauket), author of the Suffolk ban.

"They aren't biodegradable and therefore will be with us for centuries—when their intended purpose is to keep food in a package for a few minutes," said Vito A. Turso, a New York City Sanitation Department spokesman. "It seems like just something that we can do without."

But the fast-food and plastics industries say the attack on plastic packaging is a smokescreen for politicians who are ignoring more weighty garbage problems. And they say the foam is being reformulated to stop the emission of chemicals that damage the earth's ozone layer when the material is burned. They also say that claims of other dangerous gases being produced by the burning of polystyrene are unsubstantiated by scientific fact.

"You're talking about a small fraction of one percent of solid waste that is takeout food foam packaging," said Roger Bernstein of the Society of the Plastic Industry, a Washington, D.C.–based trade association. "The issue is embroiled in symbolism."

McDonald's says it has been unfairly singled out for criticism. The company says its plastic foam containers are essential for heavier foods, like big burgers and breakfasts. "For heavier food, it's more difficult to keep them hot and fresh," said Lana Ehrsam, spokeswoman for McDonald's at its Oak Brook, Illinois, headquarters. "We can't live without keeping our quality guarantee."

Both McDonald's and Dow Chemical, a major producer of materials for plastic foam, say the only environmental problems come from chlorofluorocarbons—a gas that is pumped into the plastic foam and provides its bubbly texture. Experts agree that the gas is a threat to the ozone layer, which protects people from harmful radiation from the sun, and have called for restrictions on its use. Both firms have asked their manufacturers to avoid using the gas. Dow objects to singling out Styrofoam for criticism. Styrofoam, a trademark owned by the Dow Chemical Company, is a form of polystyrene that is not actually used to make cups, plates, or fast-food packages; it is used in insulation and foam blocks for the boating and floral industries.

Some fast-food outlets do not use plastic foam containers of any kind. But Bernstein of the plastics society said the alternative food packaging methods used by other fast-food chains are no improvement on the plastic clamshells. "No one in their right mind would replace these containers with plastic-laminated paper or plastic-lined cardboard boxes," he said. "There's no advantage in that shift, because both can't be recycled once they've been contaminated by food."

With the fury of a food fight, both sides are gearing up for battle. The plastics industry has already lobbied hard for changes in the Suffolk bill. Meanwhile, state Consumer Affairs chief Richard Kessel, a proponent of a voluntary statewide ban on plastic foam containers, says he is looking for other ways to eliminate unneeded wrappers.

"[Reducing] packaging is one of the least expensive ways to help solve the garbage problem," he said.

recyclable and will pose problems in a landfill," said Allen Hershkowitz, a New York–based environmentalist. "Some will be incinerated."

If that is to happen safely and cheaply, something must be done to deal with the problems: the high costs, the questions of whether eclectic American garbage will burn efficiently without disabling European-style incinerators, the toxic chemicals in the ash.

A large part of the dilemma would be solved if the alternatives that incineration has eclipsed become reality. Source reduction and recycling can lessen the need for massive incinerators; every pound of trash that either goes uncreated or gets diverted from the garbage bin is a pound of burden lifted from the municipality faced with the multimillion-dollar investment in a burner.

Waste reduction programs can also help to address the environmental and technical questions. Glass, aluminum, and iron do not burn; their presence in incinerators lowers the burning temperature, which lessens efficiency and produces more toxic materials in the leftover ash. It also means they must be landfilled anyway, but after they have wreaked havoc in the furnace.

"Putting incombustibles into a furnace decreases the likelihood that the furnace is going to reach its necessarily high combustion [temperature]," Hershkowitz said. "To the extent metals are going in, you are increasing both the amount of pollution and the amount of final product."

Much of the technology upon which the new generation of American incinerators is based has worked because the material that should not be put into the furnace is not. "In Europe, they try to take out the components that they know would create environmental problems, like batteries," said Edward Repa, manager of solid waste disposal programs for the National Solid Wastes Management Association.

It has become increasingly apparent that household garbage in the United States contains hazardous materials—cadmium, mercury, lead—and they become concentrated in incinerator ash. Federal regulators who once exhibited a laissez-faire attitude toward incinerator ash have become more wary, warning that ash can be dangerous.

The EPA has said it will create a special category and is expected to issue guidelines for safe disposal of ash, that, for example, unmixed fly ash—the material that is caught in air pollution equipment—must be buried in a lined landfill that contains nothing else, then monitored for pollution. If fly ash is combined with the ash that comes out of the bottom of the incinerator, it can be dumped in a landfill with other garbage.

There are, however, some questions about the safety of that arrangement. No one knows what kind of poisons may leach from a landfill that contains both ash and unburned waste.

Some promising studies on reuse of ash as a raw material may help the dilemma. Frank Roethel at the State University at Stony Brook is experimenting with ash solidified into cinder blocks. So far, it seems to create a usable product that cuts pollution leakage to a minimum. The EPA is also experimenting on such uses.

But time is running out and, before the research is done, the number of incinerators—and the amount of ash they produce—are likely to increase dramatically.

Thomas Jorling, state commissioner of environmental conservation, said that he sees his department taking a bigger role in local decision-making. "We need to do more with local government in regulation, technical assistance, and planning," Jorling said. "We want to make the ground rules clearer, provide stability and confidence in the regulations."

Often lost in the arguments over incineration and recycling is the fact that none of the alternatives being discussed now will eliminate the need for landfills.

Even if recycling or burning becomes widespread, landfills will still be needed for the materials that cannot be recycled, for the ash that remains after burning, and for garbage produced during periods the incinerator is shut down.

At one time, municipal garbage landfills were not expected to pose significant environmental hazards. But many have shown up on the lists of hazardous waste sites that will take millions of dollars to clean.

There are no guarantees that today's state of the art will not be tomorrow's environmental nightmare. In the meantime, New York State's existing requirements for new landfills provide a blueprint that many in the environmental movement and the garbage industry agree can minimize the dangers:

- Landfills should not be located over areas that recharge rainwater into groundwater aquifers used as a source of drinking water.
- New landfills should be lined with a synthetic material, plus two layers of clay, which minimizes, but does not completely control, movement of contaminated water through and out of the garbage heap into the groundwater.
- For the contaminated water that slips through each liner, a set of collection pipes is required to draw it off to a place where it can be collected and treated.
- Wells must be dug to the groundwater around the landfill to monitor it for leached contamination that gets through the front-line defenses.
- A system that recovers methane—the volatile gas given off by decomposition within the landfill—should be installed after closure of the landfill to minimize air pollution and recover a valuable fuel.

Garbage has become a complicated and divisive issue, and even with a well-conceived plan, there is an element that may still be missing: public cooperation.

The best plan for picking up and processing garbage will not work unless there is a place to put it down. Community opposition is enough in many places to back politicians into paralysis on the garbage issue. It is not just a question of landfills or incinerators: In New Jersey, even recycling centers face vigorous public opposition.

Much of the solution to the garbage crunch will involve public education—enlisting the public in recycling and source reduction, including voters in the process that arrives at a solution. That, said Hershkowitz, has been the single most important key to the success of garbage management in Japan, where the process begins with primary school children being taught the subject as part of their social studies class.

Said Murray: "I was tremendously impressed with public education in Japan. I can tell you that public education is an essential part of a solid waste management strategy." The Cuomo administration's proposals recognize this, suggesting a $3 million fund for public service announcements, creation of a committee to develop an education program, and even a curriculum on solid waste management for elementary and secondary schools.

Rich Magretto, supervisor at the Rutgers University recycling center, checks the density of particles of shredded plastic bottles.

But not until some of the questions about cost, pollution, and recycling's potential are answered will the public's cooperation be won, both the industry and environmentalists agree. Municipal officials are caught between a wary public, and state and national officials who have offered only the traditional bromide: Garbage is a local problem.

Thus the increasingly complex technical questions are being answered by consultants, many of whom work for the garbage industry; the increasingly difficult job of finding sites for new facilities is going largely undone; the questions about costs and pollution are not being answered.

In 1988 the New York legislature accomplished little in easing the state's waste problems. In an effort to accelerate siting of new landfills and incinerators, Governor Cuomo proposed that state boards override local officials who persistently refuse to site needed facilities.

The idea, however, drew little support from the legislature and Cuomo dropped it. "The question of overriding local zoning is very controversial in the legislature," said Assembly Speaker Miller. "We may have to do it with a carrot and not a stick."

But regardless of disagreements, it seems clear that state and federal government will continue to struggle with these issues.

"There is a need for us to get closer to reality in terms of being helpful to municipalities in siting facilities," said State Senator Joseph L. Bruno (R–Brunswick), vice-chairman of the State Legislative Commission on Solid Waste Management. "We're thinking of what you do for a host community. How do you encourage them to site facilities?"

Among the legislature's other priorities in 1989 will be helping municipalities close landfills, providing state subsidies to build incinerators where companies are disinclined to construct them and promoting recycling, Bruno said.

And legislators in Washington also say action is overdue. "It's an interstate problem, it's no longer just a local problem," Senator Baucus said. "It's clear to me that we have to be much stronger."

CHAPTER 14

THE PROMISE OF RECYCLING

By Alvin E. Bessent and William Bunch

Seven years before the garbage barge would set sail on its six-month cruise, before Capt. Duffy St. Pierre would get enough press coverage to popularize his Cajun recipes, and before Johnny Carson suggested sending the barge's garbage cargo to Ayatollah Ruhollah Khomeini, Islip had a plan for handling its growing harvest of trash.

The idea was to force the people of Islip to sort their garbage so that bottles, cans, newspapers, and other materials could be re-sold and reused. Islip started the state's first mandatory recycling program in 1980, in an agreement with state officials that permitted the town to continue dumping garbage at its Hauppauge landfill.

Town officials coined a name for the program—Wednesdays Recyclables Are Picked Up, or WRAP. Thousands of brochures were mailed to town residents. A recycling center was created. Islip's residents were warned they could be fined for failing to separate their trash. But WRAP did not even come halfway toward reaching its goal of collecting 500 tons of recyclables a week.

And at a showdown meeting in 1983, then-Supervisor Michael LoGrande lost patience with the complaints of residents who said they did not want to sort their garbage. "I give up," he said. "I'm never going to convince you people." And he agreed to make the program voluntary.

Today, WRAP stands for something else (We Recycle America and Proudly); town officials are trying again to enforce it strictly with garbage inspectors checking out curbside cans, and collections have tripled. But Islip recycled or composted only about ten percent of its garbage in 1987, leaving the town with a long way to go to achieve its ambitious goal of 50 percent by 1990.

Islip's experience illustrates both the promise of recycling and America's failure so far to realize that promise. Government inaction, public apathy, and economic hurdles have dogged the recycling movement since it began with the first Earth Day in 1970, which gave birth to 3,000 voluntary recycling centers in just six months.

At most, ten percent of America's garbage is recycled, according to federal estimates. In 30 states that could estimate the extent of recycling, it currently amounts to an average of just six percent of garbage. That compares with aggressive recycling programs in places like Japan, where half the garbage is reused.

"Recycling holds the most potential and is getting the least attention," said Cynthia Pollock Shea of the UN-funded World-watch Institute. "It's a departure from past practices. Right now, municipalities get a contract bid from a company like Browning-Ferris, which promises to get rid of the waste in one shot by hauling it away. But with recycling, you have to get the public to participate."

Experts believe that the shortage of space in landfills, the high cost of trucking garbage long distances and of building incinerators to burn it, and the publicity over the Islip garbage barge are likely to give recycling a major boost in the next several years. Yet some experts warn that the current push to build incinerators may lead to plants so large that their appetite for garbage will preclude recycling. The big question is whether recycling can become so ingrained in American lifestyles, government, and the economy that it makes a substantial difference.

Among the roadblocks to recycling:

- Government has not committed money and people to the recycling effort. While state governments list recycling more often than any other approach as crucial to their waste disposal plans, they have spent more than $305 million on promoting garbage incinerators and less than $8 million on recycling.

- The public has not enthusiastically embraced the idea. With little prodding by public officials, few homeowners have felt compelled to make the effort to separate newspapers, bottles, and cans from their curbside trash in voluntary programs. "I can't imagine my constituents separating their waste into four or five separate containers," said State Senator Joseph Bruno (R–Brunswick) at a recent legislative hearing. "I don't believe that's practical in New York State."

- The economy has been slow to adjust to the prospect of reusing massive amounts of household trash. Radical swings in the prices paid for materials have discouraged some recycling programs. If recycling succeeds, some officials fear that there will not be steady markets for the paper, glass, and metals collected. They point to problems like those encountered in Riverhead Town, where the price for scrap metal has dropped tenfold, about 70 tons of refrigerators and washing machines have piled up in the town landfill, and a contractor stopped picking up cardboard in 1985.

- The construction of scores of large garbage incinerators has stirred fears that recyclable items like paper will be burned instead to produce enough electricity to pay off the plants' bonds. Garden State Paper, the nation's largest newspaper recycler, is pressing the federal government to issue guidelines mandated by Congress in 1984 aimed at limiting the size of municipal incinerators so that recycling will not be discouraged.

While many Americans had not heard of recycling before the environmental movement took off in 1970, the tradition of recovering wastepaper and scrap metal in this country dates back centuries. A mill used linen and cotton rags to make paper in Philadelphia in 1690. Today, recycling is an accepted fact of life in the paper and aluminum industries, among others.

Now, there's considerable debate about how much household waste can be recycled. Some environmental groups are saying that 90 percent of garbage can be recycled or composted—not only preventing pollution but saving tax dollars as well. But New York State in 1987 set what many experts say is a more realistic, yet still lofty, recycling goal of 40 percent by 1997.

A New Life

Typical processes for recycling bottles, cans and newspapers

Bottles and cans

Homeowner sorts glass bottles and cans from household trash.

Bottles and cans are taken to intermediate processing facilities for sorting.

Bottles and cans are placed on conveyer belts.

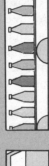

Aluminum cans are blown off the conveyer belt or picked off by hand.

Bottles are picked off the belt by hand and separated by color — green, amber and clear. Generally, one person is responsible for each color.

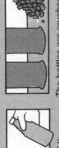

The bottles are crushed and the crushed glass is shipped to glass-container manufacturers. The glass is melted and made into glass bottles again.

Aluminum cans are crushed and sent to aluminum smelting plants, where they are made into cans or other aluminum products again.

Magnets attach steel cans to the belt. The cans are eventually knocked off and dropped into bins.

Steel cans are crushed or shredded, then trucked to de-tinning plants, which remove the tin and leave the iron to be recycled through scrap-metal dealers.

Newspapers

Homeowner ties newspapers in bales.

Bales are taken to paper mills or de-inking plants.

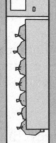

The papers are soaked and turned into a thick, lumpy liquid, 99 percent water, that washes the paper fiber free of ink.

The liquid, called slurry, is compressed between rollers to squeeze out the water.

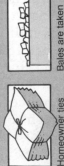

The fiber is sent to other mills, or reprocessed on the premises, to make newsprint.

NOTE: Ceramics, lead crystal, light bulbs and bottle tops cannot be recycled. Plastic bottles cannot be reused as bottles, but can be shredded and used in carpet backing, tape and construction materials.

NOTE: Newsprint is one of three grades of paper that can be recycled. The others, which must be processed separately, are high-grade paper, such as stationery and computer cards, and mixed paper, such as glossy magazine stock, egg cartons and cardboard.

PRINCIPAL SOURCES: Association for New Jersey Recyclers, "Science and Technology Illustrated"

"Historically, the economic incentive is the value of the materials," said Paul Connett, a professor at St. Lawrence University and an opponent of mass-burn incineration. "Now, another economic imperative is saving tipping fees and haulage. It's not what you make from recycling—it's what you save from recycling." In some parts of the country, taxpayers are paying the bill for shipping garbage at a cost of more than $100 a ton.

The Environmental Defense Fund, in a 1985 study of five trash incinerators planned for New York City, said that both the incinerators and an aggressive recycling program could extend the life of city landfills by seven years, but burning would cost an estimated $26 to $32 a ton, while recycling would cost $15 to $20 a ton.

Much of the new push for recycling of household wastes is based on the experience of other societies, such as Western Europe and Japan, where the practice is widespread.

While the U.S. production of garbage continues to rise, in Japan it has actually remained about the same from 1976 to 1986, with residents recycling an estimated 50 percent of their wastes. Author Allen Hershkowitz found that in Machida, one of Japan's most successful recycling cities with as much as 65 percent of materials recovered, "Public officials and solid-waste management workers go door-to-door at least once a year explaining the waste disposal system and the benefits of waste separation and recycling."

In West Germany, where the Green Party and other groups have promoted aggressive recycling programs, residents deposit their glass bottles in more than 34,000 bell-shaped "bottle banks" placed centrally in neighborhoods, and some homeowners are charged for garbage collection based on how much they throw away.

Most U.S. recycling programs have not fared as well, although officials in towns such as Islip say they are now learning from past mistakes.

Thomas Hroncich, who was Islip's environmental control commissioner when the WRAP recycling program began, said that after a 1983 flap over fining homeowners, LoGrande "ordered me not to enforce the ordinance because people would have been very upset."

AROUND-THE-WORLD PAPER TRAIL

By Ron Davis

The newspaper you throw out today could be part of your new Hyundai next year.

In 1986, more than 1.67 billion tons of wastepaper were shipped out of the Port of New York. But it ·may not be gone forever.

The story of goodbye and hello is told dozens of times monthly in New York and other American ports: The paper trash is shipped to foreign nations—where trees needed to make paper are more scarce—and used in the production and packaging of goods, many of which are shipped back and sold in the United States.

The waste—old newspapers, cardboard, and office refuse—follows a paper trail to Taiwan, South Korea, Mexico, Spain, and mainland China, where its wood fiber is used to manufacture such items as facial tissues, cardboard boxes for television sets, and interior panels for the trunks, doors, and glove compartments of automobiles.

In New York City, wastepaper is by far the leading export cargo by volume. In 1986, the Port of New York exported 747,494 long tons (2,240 pounds per ton) of wastepaper. That is more than 1.67 billion pounds of paper, worth about $80 million, according to the Port Authority of New York and New Jersey. Plastics were a distant second, with 146,940 long tons, and textile waste was third, with 81,938 long tons.

William E. Hancock, a manager at the Manhattan offices of the American Paper Institute, estimated 1987's wastepaper exports at 4.2 million tons, a slight increase over the 3.8 million tons shipped abroad in 1986.

He added that the exported amount reflects only a fraction of the almost 50 million tons of wastepaper generated in the country. Americans are expected to recycle more than 18 million tons of wastepaper this year.

"There's a trade imbalance" existing between America and other industrial nations, said Port Authority spokesman William Cahill. "Ships come in with containers, which are then unloaded. Rather than send empty ones back, we ship wastepaper."

Some of these other nations have practiced recycling almost to an art, using their own wastepaper over and over again. But this very practice, combined with limited sources of raw materials, make it necessary to look beyond their borders for sources of fiber.

"Europe and Asia are faced with the problem of recycling their own stuff so much that the wood fiber

William Wilson loads papers at the R2B2 recycling plant in the Bronx.

becomes weak, so they buy our wastepaper to get the newer, stronger fiber," Hancock said.

Taiwan was the biggest consumer of American wastepaper in 1986 with 892,000 tons, according to the U.S. Bureau of Customs. It was followed by South Korea with 839,000 tons, and Mexico, Japan, Canada, Italy, and Spain with less than 600,000 tons each.

At a Bronx transfer station called R2B2—shorthand for Recoverable Resources/Boro Bronx 2000 Inc., a subsidiary of a nonprofit development corporation—six to eight truck containers of wastepaper, or about 1,300 tons, are shipped out monthly, according to David Muchnick, president of the recycling firm. "All of it is exported out of the metropolitan area, and much of it goes overseas to Spain and Korea," Muchnick said.

Muchnick estimates that each ton of recycled paper spares 17 "average" trees from being harvested for use in paper production. "I don't know how big an average tree is," he said, "but anything recycling can do to save trees should be welcomed."

RECYCLING ROLLS ALONG

By Ford Fessenden

Late in 1986, Pennsauken, New Jersey, sent its garbage collectors out to every house in town to drop off small buckets and flyers telling residents the days of carefree discarding were over—they were going to have to start recycling.

One of Florence Johnson's neighbors did not seem to take it seriously. The bucket soon appeared in the backyard, holding clothespins.

But up and down Remington Street in working-class Pennsauken, just about everybody else started putting their buckets out, filled with cans and bottles, along with bundles of newspapers, for the Tuesday pickup by the town.

"We like it," Johnson said. "On the whole, if you go up and down our block, everyone seems to have something put out."

Camden County is testing one of the more controversial hypotheses of the garbage crunch—that you can turn the American discarder into a recycler. While a debate rages between environmentalists and the garbage industry over how much recycling the American consumer will stand for, Camden County is proving that you can do a lot—if you get out on the street and tell people about it, and if you make it easy.

When Camden's main landfill in an adjoining county closed in 1984, the county freeholders decided to try recycling to cut down the amount of garbage they had to ship out of state. Rather than rely entirely on the public spirit of citizens, the freeholders made it mandatory: Every town had to recycle 25 percent of its garbage.

Although the towns have done little more to enforce the law than post warnings on the doors of recalcitrant discarders, the notion has caught on.

The program seems to have worked better than Islip's, which was not heavily promoted and did not include town-provided containers for cans and bottles when it was started. When Islip began promoting the effort, and gave residents recycling containers, participation tripled.

Since April, 1986, when a processing center for containers was finished, Camden County has seen its recycling percentage rise steadily to 22 percent of residential waste in 1987, and it is not expected to stop there.

"We can do more than we're doing," said county recycling coordinator Jack Sworaski. "The state has a goal of 25 percent now, and we have a personal goal of 35 percent."

Some of the suburban towns

that started in 1985 have as many as 85 percent of their residents participating. In the central city of Camden, the recycling effort began in January, 1987. Only two tons came in that month, a tiny fraction of the garbage of 85,000 people. But since then, the total has grown, and Sworaski says 30 percent of residents are now participating.

"It's not an overnight thing," he said.

The key in Camden has been the county's ability to make recycling almost as painless as discarding. Residents must separate paper from the garbage and other recyclables, but they do not have to put tin cans, aluminum cans, and bottles into separate bins. They are all thrown into one container, which is picked up by the towns.

"I'm so used to it, it's easy," said Ellen Hall, 27, a neighbor down the street from Johnson who had

In Camden County, N.J., a vehicle moves cans to a truck as part of the recycling process.

Newsday/Audrey C. Tiernan

never considered the notion of recycling before she got her bucket.
"It was easy to adjust to, and it's no
problem at all."

The system requires someone to
separate the glass from the metal
containers, and that is where Resource Recovery Systems of Old
Lyme, Connecticut, comes in. In
the weedy yard of a defunct scrap
metal dealer, the company's gritty
assembly line processes up to 80
tons of cans and bottles a day. A
magnet yanks out the tin-plated
steel cans; a fan blows the light-
weight aluminum cans onto another conveyor belt; a half-dozen
employees, gloved hands a blur,
toss glass into three bins, separating by color. At the end of the belt,
a small stream of plastic and broken glass heads for a truck to the
landfill.

The operation is noisy, messy—
and profitable. "We've been running at a profit of $15,000 to
$20,000 a month," Sworaski said.

"The way I look at it, I'm not
competing with other recyclers,
I'm competing with raw materials," said Peter Karter, president
of Resource Recovery. "We pro-

LoGrande said he decided to stress public education rather
than strict enforcement of recycling because he did not want to
punish residents for a garbage crisis that he felt was the fault of
state government. "I found it despicable to give summonses to
families when the failings came out of Albany," he said.

Elizabeth C. Gallagher, executive assistant in the town Department of Environmental Control, said, however, that the early
recycling efforts in the town did not place enough emphasis on
educating homeowners. In addition to a program that rewards
recyclers with free restaurant dinners courtesy of a local radio
station, Islip is seeking to educate children about recycling with a
curriculum to be distributed in the town's school districts, poster
contests, and tours for youth groups.

The push to educate the public to recycle has had mixed results. Islip officials said that 24 percent of the material collected
in 1987 through WRAP was landfilled—partly because of staffing
problems but partly because homeowners threw in materials
unusable by industry.

"It's clearly one of the problems," Gallagher said. "We say glass
bottles and jars, they give us plastic. . . . They have good intentions. But it can be reused only when we have a market."

Some residents who participate in town recycling programs are
not enthusiastic.

duce a high-quality material, and we can find buyers."

The profits are shared with the county and towns, but Sworaski doesn't emphasize them. More important, he said, is the money the towns have saved by not paying tipping fees—which also helps convince them to enforce the recycling requirement with their citizens.

"Most towns have realized the benefits of compliance," he said. "Tipping fees have gone from $2 a yard in 1980 to an average of $43 now."

Countywide, nearly two-thirds of the households are recycling garbage, and Sworaski hopes to swell that further by enlisting commercial businesses that throw out large amounts of recyclable paper and cans.

"We've proven you can do much more than 10 or 15 percent," Sworaski said. "If the dedication is there, if the people who are promoting it are really enthusiastic supporters, and, above all, if they make it easy for the people, it works."

West Sayville resident Nancy Van Essendelft, who was cited by Islip in 1982 for not removing labels from the cans she put out for recycling, said she now removes the labels, fills a plastic pail with cans, bottles, and newspapers, and brings it to her curb every Wednesday.

"I don't really think this recycling is going to amount to a hill of beans as far as cutting the amount of garbage, but it's the law, and I certainly don't want to be caught again," Van Essendelft said. "Technically, it's a pain in the neck."

Yet Islip is hoping that its residents, along with a push to require businesses to recycle that started in January, 1988, can recycle 50 percent of town trash, because the town's 500-ton-a-day incinerator can handle only half of Islip's garbage flow. Yet some are skeptical; LoGrande said he would be surprised if Islip surpasses 30 percent.

In New York City, recycling has been slow to catch on. The city Sanitation Department recycles only 21,000 tons of trash a year—less than one percent of the trash it collects and about the amount the city heaps onto the mountains of the Fresh Kills landfill in a single day. Only after it approved a plan in 1985 for its first incinerator at the Brooklyn Navy Yard did the city create a recycling office; the $10.2 million program is a mere fraction of the total sanitation budget of $725 million.

GLOBAL VIEWS OF GARBAGE HANDLING

By Irene Virag

If Americans want to make the most of their garbage, they might do well to take their heads out of the landfill and look around the world at the new uses other countries have found for old discards.

Bonn burghers dump used car oil in public collection barrels, the Japanese toss used batteries into special retrieval boxes, and the Swiss find ways to reuse 46 percent of their glass.

In other places, people have learned what some Americans are just discovering—almost everything has its use or reuse. Rotting food can become fertilizer. Metals can be reclaimed for industrial uses. Garbage can be used to build islands. Mexico's government publishes a 106-page book entitled "Manual of Garbage and Art" with tips on how to turn tin cans into toys, and old tires into tables.

The most organized recycling programs are taking place in the Orient and in Western Europe.

Japan, where 122 million people live in an island country about the size of Montana, recycles about 50 percent of its trash. In Tokyo, residents divide garbage into specific categories—combustible, non-combustible, pollution-causing, recyclable. Anything that cannot be given a new lease on life disappears into a computerized processing system that sorts and reclaims useful materials like metals, crushes what it can, then burns what is left. Different materials are collected on different days. In some places, the garbage trucks entertain neighborhoods by playing music.

In West Germany, citizens of Bonn receive an annual 40-page illustrated guide to garbage, describing more than 100 varieties of trash from frying fat to dirty diapers to dead dogs—and what to do with them. The city provides

---

Like many cities, New York made a push toward recycling in the early 1970s by installing racks for newspapers on its trucks. But it found residents put out more bundles than the trucks could collect. Newspapers were left behind or strewn in the streets, and the program faded. Some officials now believe that New Yorkers—with a high percentage of apartment dwellers who simply toss their trash down chutes in their buildings—will be particularly tough to sell on separating garbage.

But city officials are encouraged so far by the success of a pilot

270 special communal depositories for glass, barrels for used motor oil, and a telephone hot line for citizens in need of advice on the correct disposal methods for old medicines, tires, and thermometers. The city even has a collection point for the corpses of small animals.

The Bonn garbage guide also counsels citizens against plastic bags and cellophane-wrapped vegetables and advises them to compost biodegradables and to use rubber plungers instead of caustic chemicals to clear blocked drains. "We know today that affluence cannot be equated with quality of life if reckless handling of waste . . . so burdens our environment to the point that even our health is threatened," the guide says.

If an economy of plenty and a subsequent glut of trash lead to the necessity of being earnest about garbage, poverty can also be a goad to a waste-not philosophy. People in underdeveloped nations reuse garbage out of need.

In places such as Manila and Jakarta, Indonesia, scavengers find their ways of life in garbage dumps. In Calcutta, India, animal bones are collected and boiled for use as fertilizer, and people wash used coal so they can sell what is left because it can still be burned. And in Shanghai, China, residential garbage is carted to agricultural communes where it is combined in a methane-producing pit with weeds, rice stalks, and animal and human waste. A system of pipes connects the so-called digester with little ring burners in nearby kitchens. A man turns the spigot, strikes a match, and puts on the kettle.

Following is a look at two places that handle garbage in distinctly different ways.

curbside recycling program launched in six neighborhoods. In the Forest Hills–Rego Park section of Queens, the Sanitation Department expected to collect 4.8 tons of newspapers each week, but instead the haul has ranged from 11 to 13 tons.

"They know there's a crisis. They know something has to be done—it won't be solved by magic," said Laura Denman, the curbside project's manager. "A lot of people have a feeling of uneasiness about the amount of waste in America—the throwaway society. This is something they can do."

There have been U.S. success stories. One is Camden County, New Jersey, where 22 percent of garbage is recycled, at a profit, less than five years into a mandatory program. Aided by a recycling plant where metal and aluminum cans and bottles are sorted with the aid of magnets, fans, and human hands, county officials have set their sights on recovering 35 percent of the trash.

In Rockford, Illinois, the approach has been low-tech. In Rockford's "Trash for Cash" program, a 24-year-old advertising executive who bills himself as "Trashman" dresses like a clown, drives a polka-dotted garbage truck, and hands out $1,000 prizes to recyclers. Collections of newspaper have tripled.

While Long Island's recycling programs lack the flair of Rockford's "Trashman," most town officials say local efforts have picked up steam—largely because of the publicity surrounding the garbage barge.

"With the notoriety of the barge, garbage washing up on the shore, and pollution on Long Island, people are so saturated with the news, they realize the dilemma," said Alan Sanchez, the recycling coordinator in Smithtown. The mandatory program there has attracted the participation of more than 40 percent of the households in the neighborhoods involved so far.

A Media-General/Associated Press poll in 1987 indicated that incentives for recycling—be they laws like Camden County's mandatory program or deposits to encourage return—can convert discarders into recyclers. In states that have mandatory deposits on beverage bottles, three-quarters said they routinely recycled cans and bottles; most of those said they would not if there were no deposits.

For recycling to succeed, government needs not only to pass laws but also to back them up with money, according to proponents. Islip had hoped to bolster its recycling program through a $152,000 matching grant from New York State for new equipment to help sort and process recyclables. The town applied in 1983, but the grant was not approved until 1985, and the money did not arrive until November of 1987—a two-and-a-half-year lag.

"I would call the State of New York every Monday morning and ask them where my money was," Islip's Gallagher said. "Every Monday since May, 1985, that was my job."

While the state and federal governments have paid lip service to recycling, more money and effort has been committed to incinerators.

The *Newsday* survey of 50 states, 5 territories, and the District of Columbia found that recycling is mentioned as a key objective in the solid waste disposal plans of 30 states—more than any other option. Yet state governments reported spending 39 times more money on garbage incinerators than on recycling.

"Often the funds allocated to the goal of recycling aren't there," said Shea of the Worldwatch Institute. "If you say the goal is 50 percent but only 5 percent of the budget is for recycling, then you are never going to do it."

In New York, even though 83 landfills closed in 1985 and 1986 and Long Island towns have trucked their garbage as far as Kentucky, the state Department of Environmental Conservation did not have a special recycling program until early 1987, when the agency also released a waste management plan sought by the legislature since 1980.

"There was not an aggressive recycling program at DEC until 1987," said Evan Liblit, the acting chief of the agency's bureau of source reduction and recycling. Liblit said that $167 million in state money has been committed to incinerators and only $3.2 million to recycling. He also said the DEC is developing a how-to-do-it book on recycling that was to have been distributed in February, 1987, but the material was out-of-date because recycling is changing rapidly, and there were no staffers who had time to edit it. The book is being redone.

Liblit noted, however, that 60 percent of the paper purchased by the state government in New York is recycled paper. And the Empire State is also one of seven states with mandatory deposits on beer and soft-drink cans and bottles.

Congress ordered federal agencies in 1976 to begin using recycled materials, a first step in fostering markets in the private sector. The federal Environmental Protection Agency ignored its mandate to develop guidelines. In 1980 and again in 1984, Congress ordered deadlines for issuing the guidelines, which the EPA missed. In October, 1987, draft guidelines for the use of recycled paper were finally published, but they added little incentive, saying only that recycled paper should be used if it costs the same as virgin paper.

And procurement guidelines for recycled paper that states including Michigan, Maryland, California, Florida, and Missouri have enacted following the ostensible federal mandate in the 1970s have "fallen by the wayside," according to Jere Sellers of the research firm Franklin Associates.

A. James Barnes, EPA deputy administrator, acknowledged that the agency, stung by criticism that it has done little on source reduction or recycling, is looking at its role. Said Barnes: "Certainly one of the areas we want to have on the platter is to look at the relationship between federal procurement and the beneficial effect it can have on increasing or supporting markets" for recycled goods.

Municipal guarantees of a minimum flow of trash to incinerators—a contractual arrangement that is crucial to the investment bankers who arrange financing for the plants—can harm recycling. A report prepared for New York's Legislative Commission on Solid Waste Management said that flow-control laws discriminate against recycling programs by channeling all the garbage into incinerators. Bill Kovacs, a Washington, D.C., environmental lawyer representing Garden State Paper, said that when the firm seeks to set up recycling centers in communities that are planning to build incinerators, it continually "comes into direct conflict with investment bankers."

Some planned incinerators can handle more garbage than the community produces. In Hempstead, for example, the plant will burn at least 750,000 tons per year, while the town expects to generate 640,000 tons of burnable garbage. Thomas Gulotta, the Nassau County executive and former Hempstead presiding supervisor, recently called that a margin of safety so that Hempstead can help other towns when incinerators are down for repairs, and vice versa. But some experts fear such large plants will take away the incentive for recycling.

"As the incinerators begin to demand more paper refuse, it is a disincentive to recycle paper on a municipal level," said John McCormick of the Greenpeace environmental group. "There are examples where the recycling program was directly at odds with the need to feed the demands of the resource recovery plant."

But Harvey Alter, manager of resources policy for the U.S. Chamber of Commerce, said recycling actually can benefit a garbage incinerator by removing materials such as glass and metals that do not burn. "The so-called high-tech plant can exist and does exist side by side with recycling," he said.

With little action on the government level and spotty participation by homeowners, the commercial markets for recycled products have been unreliable. "There are no fixed markets for this stuff," said Bert Cunningham, the deputy supervisor of North Hempstead, where the town is now recycling newspapers, glass, and metals that amount to about 7 to 8 percent of its waste stream.

North Hempstead used to separate about 40 to 50 tons of ferrous metals a week and sell them to a firm in New Cassel. But since that firm went out of business in the summer of 1987, the town has been unable to market the material. It stored a week's worth in roll-off containers, then dumped it in the landfill. Now, it does not separate the metals.

In some industries, the new wave of town recycling programs may compete with long-standing, little publicized, commercial recycling of scrap metal and wastepaper that has been profitable for decades.

In the wastepaper industry, collections have nearly doubled since 1970, when 12.1 million tons were used domestically and 400,000 tons were shipped overseas. In 1987, the industry used 19 million tons here and shipped 4.2 million tons abroad. And more scrap metal—11.5 million tons—was shipped to foreign nations in 1986.

"All our Chevys go to Japan and become Toyotas," said Gordon Boyd, executive director of the New York legislative commission. He said that New York commercially recycles 6 million tons per year of metal, paper, and glass bottles and that government will have to promote new, domestic markets if recycling of household trash is to help ease the garbage-disposal crisis.

Boyd said the capacity of mills in the New England–New York–New Jersey area that use wastepaper is slated to increase by only 180,000 tons by 1989. That would handle one percent of the New York State trash that is now dumped or burned.

Some officials, like Bob Hunt, vice president of Franklin Associates, fear that the push to recycle will flood markets and lower prices. Said Hunt: "I don't think there's any doubt that the recycling industry has some real rough days ahead."

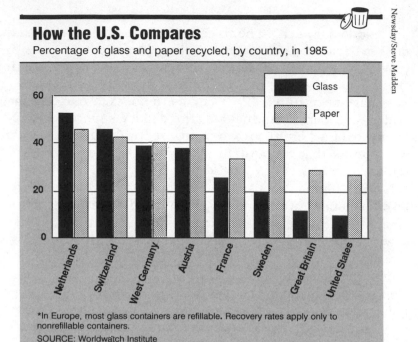

How the U.S. Compares

Percentage of glass and paper recycled, by country, in 1985

Glass

Paper

60

40

20

0

Netherlands
Switzerland
West Germany
Austria
France
Sweden
Great Britain
United States

*In Europe, most glass containers are refillable. Recovery rates apply only to nonrefillable containers.

SOURCE: Worldwatch Institute

But others feel that the increased supply of recycled paper, metals, and glass will encourage more factories to reuse materials instead of buying raw stock, and thus expand the markets. Said Brookhaven Town's recycling director, Richard Roznoy: "When business knows this is long-term and not a passing fad like recycling in the early 1970s, it will make more of a commitment to it. It was sort of a hippie thing to do. It's gone beyond that now."

Some towns are seeing signs that recycling can work, with the right approach.

In a novel experiment in East Hampton, the town had 100 families separate their trash into four containers: one for bottles and cans, one for food wastes, one for paper products such as newspapers, and one for all the garbage that cannot be recycled. To the surprise of even the most optimistic officials, separated recyclables accounted for 85 percent of the trash. The town is barging its containers and papers to a Groton, Connecticut, recycling plant.

"Politicians are the first ones to say the public will never go for recycling," said Pat Trunzo, an East Hampton Town councilman, "but, A, they never asked them, and B, they never gave them a chance."

MEXICO: A CITY DUMP IS HOME

By Jim Mulvaney

For more than 16,000 men, women, and children in the teeming city of Mexico City, garbage is literally a way of life.

They are the *pepenadores*, the collectors, who live in cardboard villages atop Santa Catarina, a 150-acre dump. They eke out a living by picking through the mounds and recovering any bit of paper, metal, or plastic that can be sold to a recycler.

And they do it under the direction of crime bosses, who control the dump and much of the recycling of garbage in this city, where about 20 million people produce a total of more than 12,000 tons of trash a day.

The city operates 2,000 "garbage trucks," mostly open-bucket dump trucks that spill much of their loads on the cobblestone and pot-holed streets. In the middle-class and rich neighborhoods, the twice- or thrice-weekly coming of the truck is announced by the clanging of a cow bell. Maids appear with plastic trash bags and the *mordida*. This is Spanish for "bite" and is the local term for bribe—in this case a payment of up to 5,000 pesos (about $2.25) for each collection. The garbage collectors rely on the *mordida* to supplement their minimum wage of about $3.50 a day, barely enough to support a family.

In poor neighborhoods and middle-class apartment districts, there is no *mordida*, but before dawn, teenage street gangs rip open the sacks for recyclables. Recyclable material is anything that can be resold—bits of plastic sheeting, tin cans, bottles, broken appliances, or hunks of metal. The teenagers are controlled by local crime bosses, who also run recycling shops. The gangs are not permitted to sell to anyone else.

At dawn, the garbage trucks collect the trash and bring it to Santa Catarina in the eastern section of the city on the site of what was a huge lake. The *pepenadores* pick through the loads, searching for the items the crime syndicate has given each of them permission to collect. The men go after the most valuable commodities, mostly metals; old women and children look for small pieces of plastic and paper.

In late afternoon, trucks from the syndicate buy the material by the ton. The *pepenadores* quietly claim that they are cheated on each load. But they say they make a bit more than the $3.50-a-day minimum wage.

"I have lived here since I was ten," said Ramona, a 30-year-old mother of six, who declined to give her last name. "I don't even notice the garbage."

"If I worked in a factory, I would have a boss who would shout at me," said Jaime Fernandez, who has lived in the Santa Catarina dump for 40 of his 45 years. "Here I work the hours that I choose. I am a free man."

The dump is a steamy place with little topsoil covering the rotting, fly-infested trash. There is no running water, nor even outhouses, and the mounds are etched with rivulets of fresh, raw sewage. The shacks, built from cardboard and metal sheets and the fenders, doors, roofs, trunks, and hoods of old cars, all have electricity. Most have black-and-white televisions. Meals are prepared on makeshift burners connected to portable gas tanks.

"In the newspapers and television, they say we live like pigs. But that's not true," said Juan Ortiz,

A child amid the shacks at the Santa Catarina garbage dump in Mexico City.

Damian Dovarganes

16. "Look, we bathe here. We wear clean clothes. We go to parties. You wouldn't recognize me if you saw me going to a party."

The children get most of their clothes from the trash, and some scurry about half-naked, helping to search for salable items or anything that passes for a toy. Most start school, but few get very far.

The city delivers water in tank trucks every day. But the filth is inescapable. Skin disease is chronic, as are lung and stomach problems. "They say we eat dead cats," said Raul Garcia, 26. "It is not true. We buy food at the markets like everyone else." However, many *pepenadores* admit they rarely pass up the remnants of a quart of milk turning solid or relatively intact chicken carcass. "We have stomachs of iron," one said. "Of course we have diarrhea, but doesn't everybody?"

City officials defend the system, claiming that the landfill would fill up too rapidly without the *pepenadores* and that the system gives work to people who would otherwise be unemployable.

For a quarter-century, the *pepenadores'* employer was Rafael Gutierrez Moreno, who controlled the garbage empire through monopoly, strong-arm tactics, and government blessing. Gutierrez, who was murdered in February,

1987, as a result of a family dispute, had 14 wives, countless girlfriends, Swiss bank accounts, and a front tooth encrusted with three diamonds. The 39-year-old millionaire lived in a palatial home complete with satellite dish and helicopter pad.

He owned the loyalty of the *pepenadores* through fear and benevolence. He employed armed guards at the dump but hired doctors for periodic visits and paid hospital bills for special cases. Once a year, he bused the scavengers to short, expenses-paid, camp-out holidays in Acapulco. Although the government has offered to relocate them, few have opted for the one-room apartments in sections of the city where they would be unable to find any work.

Since Gutierrez's death, one of the pretenders to his throne has proposed turning the Santa Catarina operation into a cooperative, promising to share all the profits with the workers. But the *pepenadores* are suspicious. "We don't want a cooperative," said Tomas Romero, 35. "Maybe if I were sick they wouldn't let me work. No, it is better this way. If I want to work, I work. Let us stay here. Just send us more garbage so we can make more money."

JAPAN: EFFICIENT APPROACH TO TRASH

By William Sexton

Japan's capital, Tokyo, is a metropolis that is literally gaining ground. Some of the region's most valuable real estate, including Tokyo Disneyland, used to be garbage, and more land is created daily in shallow Tokyo Bay.

Although they are not without problems—including pollution—the Japanese are meeting the challenge with sophisticated landfills, multi-use incinerators, and large-scale recycling.

Tokyo provides a graphic example. The central city holds a permanent population of 8.32 million persons that swells to 10.8 million during working hours. Three times a week, 3,500 mini-compactor trucks pick up each household's garbage, once a week the noncombustibles, and twice a month the old TV sets and the like. The total haul, plus sewage sludge that the Bureau of Public Cleansing also must dispose of, came to 4.95 million tons in fiscal 1985–86.

The key to getting rid of it all is sophisticated landfill design, walled in with deep steel piling, bedded on clay, covered with layers of sand, and equipped with piped sewage systems to draw off and treat the drainage. Tokyo's first municipal golf course is being built on an island of garbage in Tokyo Bay. And swimmers and windsurfers flock each summer to the shallows surrounding an earlier garbage island, now a public park reachable by ferry.

And although it has not always been easy, city fathers have successfully persuaded heavily populated areas—suburbs, particularly—to accept huge incinerators right in the neighborhood. The latest model of the Japanese incinerator comes with attached warmwater swimming pools and air-conditioned civic centers, all energized by garbage-power and—according to Taburo Kato, until 1987 chief of the Health Ministry's division of waste management—architecturally "just like museums, like art galleries, beautiful to look at."

If Tokyo has not exactly found a gold mine in garbage, its ambitious waste-to-energy program offers a silver lining. In fiscal 1985–86, municipal incinerators earned $9.2 million from sales of surplus electricity and heat to public utilities. That is in addition to supplying power to neighborhood children's centers, facilities for the elderly, and public baths. Moreover, the new $228 million Ohta No. 2 incinerator, begun on a harborside landfill in April, 1987, for completion in 1990, will convert plastics and other potentially toxic wastes into sand-like construction material, and will recover heavy metals for resale.

The city's recycling starts at home. To help Tokyo cope with its annual four-million-ton haul of garbage, for instance, all families are required to sort their own trash into "combustible" wet garbage and "noncombustible" plastics and other potentially toxic materials. Millions more voluntarily separate paper, glass, and cans for recycling.

Garbage education starts early. Children first encounter the Bureau of Public Cleansing, as it calls itself, in Tokyo primary schools during a special segment of their social studies class. This includes field trips to incinerators and landfills. Each May, the bureau sponsors a "Town Beautification Campaign." And October is designated as "The Month for All of Us to Think About Wastes Together."

During 1987's month of thinking together, there was an open house at the city's 900-ton-per-day Suginami incinerator. Residents were entertained with an exhibition of "Amazing Items from Refuse" as well as an "Old Books Fair" and "Discarded Product Bazaar." Plus, typically, a "Seminar on Wastes for Parents and Children."

Finally, the national government accepts responsibility for safe disposal of solid waste. The Ministry of Health and Welfare and other agencies provide detailed guidelines that virtually dictate local decision-making on, for example, selection of technology for incinerators.

"They work, they're safe," says the ministry's former waste management chief, Ayako Okamoto. "I visited the sites and I know."

The most telling pressure for modernizing the 2,000-odd incineration operations in Japan, he says, is neighbors' complaints about smell and smoke. Okamoto knows about that, too. He has since moved to the government's cabinet-level Environmental Agency, where he oversees the compensation Japan pays each month to some 100,000 medical victims of air pollution.

Okamoto worries about ways to reduce the waste stream. Innovative products still go into the mass market with no consideration of their eventual impact on the environment. Two potential problems that he cites are "memory wire," a new alloy used in Japanese brassieres and now exported to the United States, and the plastic material used in diapers worn by the incontinent aged. "We are an aging society," he says. "There will be a vast amount of this [diaper material] to be dealt with. From both the sanitary and the disposal viewpoint, it ought to be thought about."

PART VIII

EPILOGUE

CHAPTER 15

IN THE BARGE'S WAKE

BY SHIRLEY E. PERLMAN

In September, 1987, a pile of ashes—all that remained of the cargo of the garbage barge—was dumped on the rising summit of a landfill in Islip. The demise of the trash was celebrated with pomp befitting a ribbon-cutting ceremony. Amens were said in offices from Long Island to Albany.

The good barge *Mobro 4000* was home in Jacksonville, Florida, awaiting a more distinguished cargo. And the people whom its six-month ordeal made famous were in other places doing other things. Capt. Duffy St. Pierre was in Baltimore docking freight ships. The tugboat *Break of Dawn* and St. Pierre's old crew were on the Gulf of Mexico towing oil-rig parts. And in his hometown of Bay Minette, Lowell Harrelson, the Alabama plumbing contractor who dreamed of a trash-to-cash shipping empire, told a reporter he was finished with the garbage business. "It's history as far as I'm concerned," he said.

But it does not matter that the trash was burned, the barge gone. Or even that federal and state investigations are under way into the financing of the whole enterprise.

The dream lives on.

With visions of money dancing in their heads—payment of more than $100 a ton to cart garbage from up north and fees of 35 cents to $5 a ton to dump it down south—opportunists on both sides of the Mason-Dixon line remain convinced that the big profits are still out there in the trash.

In Long Island City, Thomas Gesuale, president of Review Avenue Enterprises and a principal in the company that underwrote the garbage barge, is giving the dream material expression. Gesuale owns the only private waterfront transfer station in the New York area licensed to barge trash. He is building two new ocean-going barges designed to hide the garbage—steel-covered scows with an aeration system and offloading ramps.

In Alabama, Rick Byrd, a neighbor and former sidekick of Harrelson's, says he has barges, tugboats, and a landfill site in Texas and is ready to sail.

In Islip, Thomas Hroncich, president of Waste Alternatives and another founding partner of the old dream, says he will barge again with anyone who guarantees government permits for getting rid of seaborne garbage at a designated landfill.

Gesuale and Hroncich may be inveterate dreamers. Each lost $50,000 on the original garbage barge scheme. Harrelson did not invest anything. He lost only what he stood to gain.

On a dreary day, Gesuale, seated in his trailer-office at the Review Avenue yard, gazed at framed photographs of the *Mobro* teeming with trash and remembered how it all began.

The idea of a garbage barge surfaced in 1986 when the Islip landfill was closed to school and commercial trash. Waste Alternatives began trucking the refuse upstate at a charge to carters of $86 a ton. So when Lowell Harrelson appeared with his barging ideas, everyone listened.

"Harrelson came here and said, 'I can beat that price, but we have to get it down to Alabama or North Carolina, somewhere down south,' " Gesuale said. Harrelson noted that a barge the size of the *Mobro* could carry 20 truckloads.

Harrelson was introduced to the Long Island garbage scene by his friend, Rick (Hurricane) Byrd. Raised in Harrelson's hometown, Byrd had been working on Long Island with a crew of truckers, hauling trash and helping with the Hurricane Gloria cleanup. Hroncich referred to them as "the rebels."

Gesuale and Hroncich brought in four carting companies as partners—Detail Carting of Ronkonkoma, Jet Sanitation of Islip, Jamaica Ash and Rubbish Removal of Westbury, and Allied Sanitation of Queens. Each contributed $50,000. A deposit of $300,000 was made in the Farmingdale branch of Barclays Bank, and United Marine Transport Services Inc. was born.

Harrelson leased the *Break of Dawn* and *Mobro 4000* from Harvey Gulf International of Harvey, Louisiana, and contracted to dump garbage at $5 a ton in a landfill outside of New Orleans. Harvey Gulf got a cash advance of $100,000 to cover the maiden run and a series of future garbage runs from New York to the South.

The plan was to haul trash from Waste Alternatives in Islip to Review Avenue, where Gesuale would be paid $50 a ton to bale, load, and barge it. Gesuale's actual cost would be about $30 a ton, he said. Harrelson would get half of the projected $20-a-ton profit, and the six principals of United Marine Transport Services would split the other half. Harrelson would pay offloading and transport and dumping fees from his profits.

A fleet of four barges was envisioned, hauling 10,000 tons a day. The partners were projecting profits of $200,000 a day.

"He said when he first came here this was better than oil," Gesuale said of Harrelson. The barge sailed from Gesuale's dock on March 22, 1987.

Three days later, Harrelson stopped in Morehead City, North Carolina, in the hope of dumping his cargo at a landfill in nearby Jones County without having to go all the way to New Orleans. But he did not have a signed contract. And North Carolina officials ordered the *Mobro* out to sea. Nor was he able to dump in New Orleans, his next port of call. Louisiana officials said he was in violation of state health codes.

Before the story of the garbage scow was over, it had been banned from North Carolina, Louisiana, Mexico, Belize, the Bahamas, Florida, and even New York. Mexico and Cuba sent gunboats to make sure the scow stayed away.

Along the way, the barge became a symbol of the nation's garbage problem. There were barge T-shirts and cartoons, and the garbage barge made the Johnny Carson and Phil Donahue shows. One bale of the trash was preserved for Gar-Barge souvenirs—$10

Tom Gesuale, president of Review Avenue Enterprises in Long Island City, on a barge he has designed especially to haul garbage.

morsels of *Mobro* garbage in Santa Claus red packages for the holidays.

Looking wet and weary, and under police escort, the barge came home in May, 1987, after two months and 6,000 miles at sea. New York was like all the other places—St. Pierre was greeted with legal injunctions by all of the city's five boroughs.

Even though the trash finally was burned in Brooklyn and the ash buried in Islip, the controversy is not over.

Both the state Organized Crime Task Force and the FBI are said to be investigating alleged mob ties to the barge. Both agencies refused to comment on published reports that probes were under way.

Also silent on the question of investigations was Thomas Hroncich of Waste Alternatives, who has repeatedly refused to identify the consortium of carters who hold stock in his company.

Allegations of mob ties stem from a June, 1987, report by Assemblyman Maurice Hinchey (D–Kingston), who raised questions about the appropriateness of Islip Town's leasing contract with Waste Alternatives and its consortium of carters.

Two principals of the carting companies that put up $50,000 for the garbage barge—Detail Carting and Jamaica Ash and Rubbish Removal—were indicted in 1984 on charges of conspiracy and coercion in connection with a state task force investigation into mob control of Long Island carters. Thomas Ronga of Detail Carting pleaded guilty and was sentenced to probation and community service. Emedio Fazzini of Jamaica Ash was granted immunity.

But lawsuits and investigations notwithstanding, the dreamers barge on.

"If the right opportunity came along . . . yes . . . but I would want to see everything signed in blood from the governor on down," Hroncich says. "There were too many assumptions made before."

"I'd deal with anybody who can get garbage, baled garbage, on our barge and who can pay for it up front," Byrd says. He contends that he has a deal with a Texas landfill for garbage dumping. But he will not say for how much or where in Texas because he, like Harrelson, blames the *Mobro* fiasco on publicity.

Gesuale, a burly man who sounds like Rodney Dangerfield,

says he is prepared to stay on the barge ride for the long haul. That was Harrelson's mistake, he says. "He thought it would be a one-shot deal. But it's not an oil well that just comes in." Gesuale's barging operation includes two huge balers at his Long Island City site. For the time being, private parties wishing to ship trash out of New York must deal with him.

"The barge is far from over," Gesuale says. "It's just starting." Which is why he is building the barge of barges and why he is dreaming of shipping routes from New York to Texas and Long Island to Venezuela.

As for Harrelson, Gesuale says, "I feel sorry for the guy. He had a dream."

But others wonder if Harrelson has really given up the dream. "Yeah, he called me the other day with something else—another deal for a landfill," said Hroncich. And he laughed.

CHAPTER 16

THE TRASH-EXPORT BOOM

By Thomas Maier and Mark McIntyre

A year after the nation's garbage crisis floated into view—symbolized by a wandering barge of trash from New York City and Long Island—America's rush to burn its garbage in modern, multimillion-dollar incinerators has been slowed by strong local opposition, persistent mechanical problems, and regulatory delays.

In November of 1988, the state halted New York City's first waste-to-energy incinerator because of doubts about the city's recycling and ash disposal plans. The company selected to build Huntington's plant is cleaning up from an explosion at its Hartford, Connecticut, facility. Los Angeles, Seattle, Philadelphia, Massachusetts, and a suburb of Washington, D.C., have halted plans for new incinerators—causing a 10 percent drop in the once white-hot incinerator business.

"The resource recovery business has cooled off," says Albert Medioli, a vice president for Moody's Investors Service. "Everybody says that you'll still see a lot of resource recovery activity, but maybe not as much as everybody expected. Things like recycling are being seen increasingly as an option."

ᵧcling, hailed by environmentalists as the way to
rbage crisis, is encountering problems of its own
re communities have started paper recycling, the
ᵇecome glutted with wastepaper, and prices have

.on of the garbage sent to incinerators or recycling
programs anᵈ scores of local landfills shutting down, the waste
industry witnessed an extraordinary boom in the volume of garbage East Coast cities shipped as far as several hundred miles for
disposal.

Philadelphia signed a contract in the summer of 1988 to ship all
its waste beyond city borders. In the New York metropolitan area,
the public pays about $800 million annually to haul 7 million
tons to landfills outside the area.

On Long Island, citizens pay more than $69 million to ship at
least 725,000 tons of garbage and incinerator ash to Pennsylvania,
Ohio, and upstate New York landfills. These figures are low. Jet
Sanitation, a major Long Island carter, declined to disclose how
much waste it exports. Most of this money is split by shippers and
out-of-state landfill operators.

These shipments have sparked a backlash in states accepting
the trash, including a shooting in West Virginia over the dumping
of Philadelphia garbage.

As the nation sorts through the options and consequences of its
garbage dilemma, a wide range of experts believes that rising
costs and the need for environmentally superior disposal
methods will force communities to build more incinerators as
part of a disposal strategy that combines landfills, composting,
and recycling.

Builders of the incinerators, which dispose of 7 percent of the
nation's trash, are still trying to persuade elected officials that
they can handle trash in a cost-effective, environmentally sound
manner. While worries recede about some forms of air pollution,
there are new questions about the plants' role in acid rain, and the
industry is dogged by continued mechanical upsets.

In 1988, six major resource recovery plants opened in Massachusetts, Maine, Virginia, and Connecticut, according to the U.S.
Conference of Mayors. On Long Island, Babylon, Islip, and Hempstead plan to open new incinerators this year.

Most new burners meet air pollution standards, according to

Trucking It Off the Island

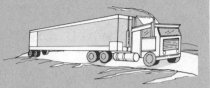

How much garbage and ash municipalities now ship off Long Island for disposal, what it costs and where it is shipped to. All the companies but Interstate Bi-Modal handle raw garbage; Interstate Bi-Modal carries incinerator ash.

Company	Municipality	Annual volume in tons	Cost in millions	Location of landfills
Waste Management	Oyster Bay	208,000	$21.4	Pennsylvania
Browning-Ferris Industries	Hempstead	414,000	39.0	New York, Pennsylvania
A-1 Carting	Parts of 13 Nassau villages*	46,800	3.9	Ohio
Interstate Bi-Modal	Glen Cove, Long Beach	55,000	4.7	Ohio

*Bayville, Centre Island, Cove Neck, Lattingtown, Matinecock, Mill Neck, Muttontown, Old Brookville, Old Westbury, Oyster Bay Cove, Roslyn Harbor, Sea Cliff and Upper Brookville.

Newsday / Linda McKenney

Marjorie Clarke, director of solid waste research at INFORM, a Manhattan-based environmental group.

"Generally speaking, the plants are meeting their requirements," she said. "They are a whole lot better than they used to be. A lot of strides have been made in reducing particulates, acid gas, and dioxin."

Reduction in dioxin, a highly toxic waste gas, has been impressive, she said. In the early 1980s, plants emitted up to 4,000 nanograms of dioxin per cubic meter of air. A nanogram is a billionth of a gram. "Now, the range is from one nanogram to twenty-five," she said.

Despite reduced air pollution, plants often fail to control nitrogen oxides, a contributor to acid rain and smog. A study of three of Ogden Martin's plants shows nitrogen oxide emissions averaged 326 parts per million. Well-run plants can achieve emissions of 100 parts per million, Clarke said. The U.S. Environmental Protection Agency has not set emission limits for these pollutants.

Besides air pollution, plant performance is still shaky. Of the nation's 111 operating incinerators, 46 percent have closed periodically for reasons other than routine maintenance.

Yet closing half the plants is better than the industry's 1986 record, when nearly three-fourths of all incinerators had un-

WHEN THE TOWNSFOLK SAID: NO MORE

By Alyssa Lenhoff

Tucker County, West Virginia, is the kind of place where people keep to themselves. Mountains and long-standing feuds separate those who live in this scenic county that borders Maryland.

"Family fights and beliefs about what the county should be have separated the people. Some want it kept scenic and natural. Others want it exploited for tourism," said Doris Cussins, mayor of Davis, one of the county's few cities.

But in the summer of 1988, the people of Tucker County came together, united in a fight against out-of-town garbage.

After reading in the local newspaper that the county commissioners—without public hearings —signed a contract with a firm to haul in New York and Pennsylvania household garbage, residents set up tents at the county landfill, determined to stop all trucks. "Who knows what's in that stuff? There could be infected needles that have AIDS on them or other diseases," resident Jerry Wilson said at the time. He was one of hundreds who manned round-the-clock pickets at the landfill.

At first, county commissioners said the trash would bring needed revenue to these West Virginia communities that used to mine coal. Much of Tucker County was wiped out during the "Great Floods" of 1985 that killed about 20, left thousands homeless and the county needing revenue to rebuild. Commissioners had said they thought they made a good deal with Byrson Associates of Bay Minette, Alabama, and they looked to the revenues as found money.

In exchange for using the landfill, Byrson agreed to pay the

planned shutdowns. Plastic gets much of the blame. When burned, it can corrode machinery in the plants, forcing costly repairs.

"There is a learning curve, but a lot of people who get into the business didn't take advantage of some of the expertise already in the field," says Andrew Martin, recent past chairman of the New York–based American Society of Mechanical Engineers' Solid Waste Division.

Two days after a $171 million Hartford incinerator began rou-

county $3 for every ton of garbage hauled in. County officials told Byrson they needed about $600,000 in a hurry. But after seeing the public's reaction, commissioners scrambled to break the contract.

At first, Greg Byrd and Rick Jackson, co-owners of Byrson, fought the opposition, and truckloads of trash began arriving June 1, 1988. Nine days later, Byrson was forced by the courts to padlock the landfill because it had accepted more trash than its permit allowed for the entire month. Opposition stepped up.

"I hate to think it will come to violence, but we didn't put up with any crap in the mines and we won't put up with it here," protester Ted Thomas said.

But violence erupted. Byrd and Jackson, who were renting a house in Tucker County, said someone fired several shots at one of their cars while it was parked at home. Police confirmed the incident.

Eventually, the public outcry proved too much for Byrson, and five weeks after its first truck brought a load of out-of-state garbage to the Tucker County landfill, Byrd and Jackson gave up. "We just don't see any point in continuing the thing when the county is not supporting us," Byrd said.

The landfill has returned to accepting only local household waste. But the people of Tucker, who once kept to themselves, still meet regularly to talk about how to keep their county free from menaces, Cussins said. "Us hillbillies are going to rise again," said Wilson, who now calls himself an environmentalist.

---

tine operation last year, an explosion tore apart steam tubes, bent boiler walls, and blew out part of the plant's power building. The operator, Combustion Engineering, which Huntington selected to build a 750-ton-per-day plant that will employ a different technology without a history of explosions, will pay $13 million to repair the damage.

Another incinerator-builder, Pennsylvania Resource Systems, ran out of money and was declared in default on a $40 million job in Islip. It will delay the Islip plant a year.

Despite industry growing pains, two leaders, Wheelabrator Technologies of New Hampshire and Ogden Martin of New Jersey, continued to grow impressively. Wheelabrator opened two major plants in the Northeast last year. Ogden Martin, which began the year with three plants, will open five more early this year.

But environmental activists assert that a future of many more incinerators is not inevitable. Several, such as environmentalist Barry Commoner, contend that recycling is a substitute for incinerators.

Commoner recently directed a project in East Hampton in which volunteer households recovered 85 percent of their waste. In North Hempstead, a citizens' group, fighting the town's proposed incinerator, has commissioned a study that found recycling would be an effective alternative to the incinerator.

Solid waste officials, however, scoff at such thinking.

"I don't think Commoner proved anything in East Hampton other than the fact that you can get 96 volunteers to do what you want to do," said Evan Liblit, the state Department of Environmental Conservation's former recycling coordinator and now Babylon's environmental control commissioner.

"You can't get [the] 15,000 people [of East Hampton] with alacrity to put all their garbage in four different cans. I don't like to see anyone put all their eggs in one basket."

Like incinerators, recycling has had a rocky road.

With more cities and towns adopting mandatory paper collection programs, a record 35 percent of all newspapers read in the United States were recovered last year, says Rodney Edwards, vice president of the American Paper Institute, an association of pulp, paper, and cardboard manufacturers.

This achievement, however, helped create a glut that overwhelmed paper mills. Last summer, prices for wastepaper nosedived, dropping from $50 a ton to $20. They since have dropped even more.

"Unless there is demand, there will always be a supply overhang in scrap markets," explained Garrett Smith, a recycling specialist with Ogden Martin.

In three years, the flood of wastepaper should ease as U.S. paper mills add $1 billion worth of new paper machines, Edwards says. "The current situation of overcollection is temporary," he said.

This year, consumers will see examples of streamlined plastic packaging. General Electric is seeking broader acceptance of a hard-plastic milk container that can be reused 100 times. Procter & Gamble plans to sell Spic-and-Span in a recycled bottle. But the nation recovers only 1.5 percent of its plastic, most of it beverage containers, said Jerry Powell, a Portland, Oregon, editor of two plastic recycling periodicals.

On Long Island, Anthony Noto, the ex-Babylon supervisor, plans to open the island's first firm to make new products from waste plastic. It will produce plastic wood for use as bulkheads, outdoor decks, and possibly snow fences. Yet, even if Noto meets his 1990 goal of using 30 million pounds of plastic, it would amount to less than 10 percent of the plastic Long Islanders discard.

With recycling still limited, landfills closing, and incinerators yet to open, the metropolitan region is spending $800 million a year, according to conservative estimates, to export waste it cannot handle locally. That's a fourfold increase from 1987's garbage-export bill.

New Jersey currently ships 54 percent of its waste out of state. In New York City, as the city doubled the fee it charges for disposal of privately collected waste, carters deliver 4,000 fewer tons each day to the Fresh Kills landfill compared with a year ago. What's not showing up, more than 1 million tons a year, is going out of state, says Mike Carpinello, director of the city's Bureau of Waste Disposal.

But when the East Coast garbage reaches its final destination, disposal can be a touchy issue.

In the summer of 1988, citizens in Tucker County, West Virginia, a poor Appalachian community, learned an Alabama firm had contracted to bring New York and Philadelphia trash to the county's landfill. Citizens protested with placards; others blocked the landfill's entrance.

One night, an unknown assailant fired several shots from a .22-caliber gun into the rear window of one of the hauler's vehicles. Eventually, the hauling firm backed off, giving in to citizen pressure.

Aside from such vigilante tactics, states burying surplus East Coast waste have moved quickly to protect themselves.

In the summer of 1988, Ohio Governor Richard F. Celeste

signed a law that authorizes counties, once they establish garbage districts, to decide if they want to take outside garbage. A federal appeals court, reviewing an Oregon law, held in 1987 that districts can refuse interstate waste.

"These states have now woken up," said Mary Sheil, deputy director of solid waste management in New Jersey. "Where will the waste go tomorrow if we are shut out?"

Despite reform of garbage rules in receiving states, coal companies and national garbage firms, seeking to cash in on East Coast waste, are preparing thousands of acres, including old Appalachian strip mines, as new landfills, garbage company officials and environmentalists say.

In rural Kentucky, where almost all people drink water from underground, converting strip mines into landfills worries Tom Fitzgerald, a lawyer with Kentucky Resources Council, an environmental advocacy group.

"Is it wise to dedicate our land for a short-term dislocation in East Coast garbage when the long-term liability is borne by our state and the people who live here?" Fitzgerald asks.

In the next five years, as incinerators open in the East and more people recycle, garbage exports will wane.

"The garbage will come back," predicts Ross Patten, vice president of American REF-FUEL, located in Houston.

In August, his firm plans to open its first plant, a 2,250-ton-per-day incinerator in Hempstead. To undercut long-haulers, REF-FUEL will charge carters from $60 to $80 a ton to take their waste at the Hempstead plant.

"We will price it so that the local option is more effective than long-haul garbage," Patten said.

In a related issue, Suffolk County's pioneering plastic law, which was passed last spring and bans non-biodegradable plastic such as grocery sacks and food wrap, has hit a snag because there is no easy way to tell whether packaging in supermarkets, delis, and restaurants contains the banned materials.

Under the Suffolk law, by July of 1989, 4,500 retail food establishments must find environmentally innocuous substitutes for polyethylene grocery bags and polystyrene or polyvinyl chloride wrapping.

To resolve what could otherwise be massive confusion, the

health department has asked the county attorney's office to help write new rules to carry out the law. While environmentalists hailed the law, plastic industry trade groups and manufacturers, seeking to overturn it, have taken the county to court.

The county is making no progress in preparing rules to implement the controversial law, according to Assistant County Attorney Fred Eisenbud. The law won't take effect for seven months and other business is more pressing, he said.

Improved waste disposal is now a top environmental priority of many states and the federal government. The federal EPA has proposed a draft garbage strategy for communities. Perhaps the toughest garbage rules in the nation took effect in New York December 31, 1988. They require new or expanded landfills to be built with four liners and two pollution collection systems. As a result, the cost of a new landfill will soar to more than $500,000 an acre.

Largely because of local opposition, almost no modern landfills have been built in New York in the 1980s.

"Those new rules can't protect the environment because no one can get one built," says Jack Baker, a lawyer for a major Kentucky coal company that seeks to open landfills in New York but cannot find sites.

The new state rules also set strict air pollution limits for incinerators. These provisions will force existing incinerators, such as in Glen Cove and Long Beach, to add costly scrubbers and state-of-the-art air pollution equipment which could raise operating costs by $10 to $20 a ton, says Floyd Hasselriis, an engineer advising a Washington, D.C., incinerator consultant.

And the rules require that 5,000 facilities handling waste in New York—infectious waste transfer stations, composting sites, waste tire and recycling centers, as well as landfills and incinerators—have state permits. It will take seven years to issue all the permits, says Norman Nosenchuck, the state's solid waste director, and landfills will require the most work.

In 1989, a new generation of garbage incinerators will open on Long Island in Hempstead, Babylon, and Islip. Pursuing a different tack, Southold is reviewing proposals from vendors that propose to compost garbage and sludge without any burning.

Nine other Long Island towns—North Hempstead and Oyster

Bay in Nassau, and Huntington, Smithtown, Brookhaven, Riverhead, Southampton, East Hampton, and Shelter Island—won't have an alternative to landfilling by December, 1990, the state deadline for ending burial of garbage on the Island to curb contamination of the Island's drinking water from underground.

Confronted with widespread non-compliance despite years of prodding, the state can grant limited waivers to towns with a clear commitment to new facilities or force taxpayers in towns with landfill space remaining in 1990 to ship garbage out of state.

Disposal of incinerator ash is another unresolved problem. As society demands that incinerators emit less air pollution, they must produce more ash. Ash amounts to 30 percent of the garbage burned in a plant.

Currently, Long Island, like many Northeast communities, has no landfill that will take ash. New York City had planned to use Fresh Kills until DEC Commissioner Thomas Jorling strongly suggested it was unacceptable. Most experts foresee reliance for five years or more on out-of-state landfills.

The Island's only incinerator ash, from Glen Cove and Long Beach, is now shipped by rail to Ohio. As three more incinerators open, more ash will go out of state. But with additional ash, the incinerator companies will experiment with facility-specific ash to identify environmentally safe alternatives.

As long as Long Island and New York City cannot open ash landfills or accept alternative uses such as building materials, road bases, or gabions to stabilize beaches, taxpayers will ship growing volumes of ash to distant landfills as more incinerators open.

Until the problems are solved, said Liblit, Babylon's environmental commissioner, "They will begin shipping their garbage into the sunset and hope it doesn't come back."

NOTE: This chapter was prepared and published a year after the original *Newsday* series.

INDEX

ABOUT THE AUTHORS

Newsday's series, "The Rush to Burn," was the result of a six-month investigation. It was published in December, 1987.

Reporter Thomas J. Maier and environmental writer Mark McIntyre began the reporting in June, 1987.

They were later joined by Alvin E. Bessent, William Bunch, Marie Cocco, Ron Davis, Walter Fee, Ford Fessenden, Richard C. Firstman, Robert Fresco, Jane Fritsch, Robert E. Kessler, Barry Meier, Shirley E. Perlman, Bob Porterfield, Michelle Slatalla, and Irene Virag. *Newsday* correspondents Jim Mulvaney in Mexico City, Adrian Peracchio in Germany, and William Sexton in Japan also contributed.

The project was supervised by Joseph Demma and Harvey Aronson under the direction of Long Island editor Richard Galant. The photographers were Audrey Tiernan and John Keating.

Patty Klein and Dorothy Guadagno were the project's researchers.

ALSO AVAILABLE FROM ISLAND PRESS

Americans Outdoors: The Report of the President's Commission
The Legacy, The Challenge, with case studies
Foreword by William K. Reilly
1987, 426 pp., Appendixes, case studies, charts
Paper: $24.95 ISBN 0-933280-36-X

The Challenge of Global Warming
Edited by Dean Edwin Abrahamson
Foreword by Senator Timothy E. Wirth
1989, 350 pp., tables, graphs, index, bibliography
Cloth: $34.95 ISBN: 0-933280-87-4
Paper: $19.95 ISBN: 0-933280-86-6

Creating Successful Communities: A Guidebook to Growth
Management Strategies
By the Conservation Foundation
1989, 350 pp., index, appendices
Cloth: $39.95 ISBN: 1-55963-015-9
Paper: $24.95 ISBN: 1-55963-014-0

Crossroads: Environmental Priorities for the Future
Edited by Peter Borrelli
1988, 352 pp., index
Cloth: $29.95 ISBN: 0-933280-68-8
Paper: $17.95 ISBN: 0-933280-67-X

Down by the River: The Impact of Federal Water Projects and
Policies on Biodiversity
By Constance E. Hunt with Verne Huser
In cooperation with The National Wildlife Federation
1988, 256 pp., illustrations, glossary, index, bibliography
Cloth: $34.95 ISBN: 0-933280-48-3
Paper: $22.95 ISBN: 0-933280-47-5

Forest and the Trees: A Guide to Excellent Forestry
By Gordon Robinson, Introduction by Michael McCloskey
1988, 272 pp., indexes, appendices, glossary, tables, figures
Cloth: $34.95 ISBN: 0-933280-41-6
Paper: $19.95 ISBN: 0-933280-40-8